Intelligent Systems Reference Library

Volume 223

The aim of this series is to publish a Reference Library, including novel advances and developments in all aspects of Intelligent Systems in an easily accessible and well structured form. The series includes reference works, handbooks, compendia, textbooks, well-structured monographs, dictionaries, and encyclopedias. It contains well integrated knowledge and current information in the field of Intelligent Systems. The series covers the theory, applications, and design methods of Intelligent Systems. Virtually all disciplines such as engineering, computer science, avionics, business, e-commerce, environment, healthcare, physics and life science are included. The list of topics spans all the areas of modern intelligent systems such as: Ambient intelligence, Computational intelligence, Social intelligence, Computational neuroscience, Artificial life, Virtual society, Cognitive systems, DNA and immunity-based systems, e-Learning and teaching, Human-centred computing and Machine ethics, Intelligent control, Intelligent data analysis, Knowledge-based paradigms, Knowledge management, Intelligent agents, Intelligent decision making, Intelligent network security, Interactive entertainment, Learning paradigms, Recommender systems, Robotics and Mechatronics including human-machine teaming, Self-organizing and adaptive systems, Soft computing including Neural systems, Fuzzy systems, Evolutionary computing and the Fusion of these paradigms, Perception and Vision, Web intelligence and Multimedia.

Indexed by SCOPUS, DBLP, zbMATH, SCImago.

All books published in the series are submitted for consideration in Web of Science.

Christos Dimitrakakis · Ronald Ortner

Decision Making Under Uncertainty and Reinforcement Learning

Theory and Algorithms

Springer

Christos Dimitrakakis
Informatique
Université de Neuchâtel
Neuchâtel, Switzerland

Ronald Ortner
Department Mathematik und
Informationstechnologie
Montanuniversität Leoben
Leoben, Austria

ISSN 1868-4394 ISSN 1868-4408 (electronic)
Intelligent Systems Reference Library
ISBN 978-3-031-07612-1 ISBN 978-3-031-07614-5 (eBook)
https://doi.org/10.1007/978-3-031-07614-5

This Springer imprint is published by the registered company Springer Nature Switzerland AG
The registered company address is: Gewerbestrasse 11, 6330 Cham, Switzerland

Preface

The purpose of this book is to collect the fundamental results for decision making under uncertainty in one place. In particular, the aim is to give a unified account of algorithms and theory for sequential decision making problems, including reinforcement learning. Starting from elementary statistical decision theory, we progress to the reinforcement learning problem and various solution methods. The end of the book focuses on the current state of the art in models and approximation algorithms.

The problem of decision making under uncertainty can be broken down into two parts. First, how do we learn about the world? This involves both the problem of *modeling our initial uncertainty* about the world and that of drawing *conclusions* from *evidence* and our initial belief. Secondly, given what we currently know about the world, how should we *decide* what to do, taking into account future events and observations that may change our conclusions?

Typically, this will involve creating long-term plans covering possible future eventualities. That is, when planning under uncertainty, we also need to take into account what possible future knowledge could be generated when implementing our plans. Intuitively, executing plans which involve trying out new things should give more information, but it is hard to tell whether this information will be beneficial. The choice between doing something which is already known to produce good results and experiment with something new is known as the exploration–exploitation dilemma, and it is at the root of the interaction between learning and planning.

Part I of the book, Chaps. 1–4, focuses on decision making under uncertainty in non-sequential settings. This includes scenarios such as hypothesis testing, where the decision maker must choose a single action given the available evidence. Most of the development is given through the prism of Bayesian inference and decision theory, where the decision maker has a subjective belief (expressed as a probability distribution) over what is true. Part II of the book, Chaps. 5–8, introduces sequential problems and the formalism of Markov decision processes. The remaining chapters are devoted to the problem of reinforcement learning, which is one of the most general sequential decision problems under uncertainty. Finally, we have added a number of theoretical and practical exercises that will hopefully aid the reader to understand the material.

Neuchâtel, Switzerland Christos Dimitrakakis
Leoben, Austria Ronald Ortner

Acknowledgements

Many thanks go to all the students of the *Decision making under uncertainty* and *Advanced topics in reinforcement learning and decision making* classes over the years for bearing with early drafts of this book. A big *thank you* goes to Nikolaos Tziortziotis, whose code is used in some of the examples in the book. Finally, thanks to Aristide Tossou and Hannes Eriksson for proof-reading various chapters. Finally, a lot of the coded examples in the book were run using the *parallel* package by Tange [1].

Reference

1. Tange, O.: Gnu parallel-the command-line power tool. USENIX Mag. **36**(1), 42–47 (2011)

Contents

Chapter 1
Introduction

1.1 Uncertainty and Probability

A lot of this book is grounded in the essential methods of probability, in particular using it to represent uncertainty. While probability is a simple mathematical construction, philosophically it has had at least three different meanings. In the classical sense, a probability distribution is a description for a truly random event. In the subjectivist sense, probability is merely an expression of our uncertainty, which is not necessarily due to randomness. Finally, in the algorithmic sense, probability is linked to the length of a program that generates a particular output.

In all cases, we are dealing with a set Ω of possible outcomes: the result of a random experiment, the underlying state of the world and the program output respectively. In all cases, we use probability to model our uncertainty over Ω.

Classical Probability
A *random experiment* is performed, with a given set Ω of possible outcomes. An example is the 2-slit experiment in physics, where a particle is generated which can go through either one of two slits. According to our current understanding of quantum mechanics, it is impossible to predict which slit the particle will go through. Herein, the set Ω consists of two possible events corresponding to the particle passing through one or the other slit.

In the 2-slit experiment, the probabilities of either event can be actually accurately calculated through quantum theory. However, which slit the particle will go through is fundamentally unpredictable. Such quantum experiments are the only ones that are currently thought of as truly random (though some people disagree about that too). Any other procedure, such as tossing a coin or casting a die, is inherently deterministic and only *appears* random due to our difficulty in predicting the outcome. That is, modelling a coin toss as a random process is usually the best approximation we can make in practice, given our uncertainty about the complex dynamics involved. This

C. Dimitrakakis and R. Ortner, *Decision Making Under Uncertainty and Reinforcement Learning*, Intelligent Systems Reference Library 223,
https://doi.org/10.1007/978-3-031-07614-5_1

gives rise to the concept of subjective probability as a general technique to model uncertainty.

Subjective Probability
Here Ω can conceptually not only describe the outcomes of some experiment, but also a set of possible *worlds* or realities. This set can be quite large and include anything imaginable. For example, it may include worlds where dragons are real. However, in practice one only cares about certain aspects of the world, such as whether in this world, you will win the lottery if you buy a ticket. We can interpret the probability of a world in Ω as our degree of belief that it corresponds to reality.

In such a setting there is an actual, true world $\omega^* \in \Omega$, which could have been set by Nature to an arbitrary value deterministically. However, we do not know which element in Ω is the true world, and the probability reflects our lack of knowledge (rather than any inherent randomness about the selection of ω^*).

No matter which view we espouse, we must always take into account our uncertainty when making decisions. When the problem we are dealing with is sequential, we are taking actions, obtaining new observations, and then taking further actions. As we gather more information, we learn more about the world. However, the things we learn about depend on what actions we take. For example, if we always take the same route to work, then we learn how much time this route takes on different days and times of the week. However, we don't obtain information about the time other routes take. So, we potentially miss out on better choices than the one we follow usually. This phenomenon gives rise to the so-called exploration-exploitation trade-off.

1.2 The Exploration–Exploitation Trade-Off

Consider the problem of selecting a restaurant to go to during a vacation. The best restaurant you have found so far was *Les Epinards*. The food there is usually to your taste and satisfactory. However, a well-known recommendations website suggests that *King's Arm* is really good! It is tempting to try it out. But there is the risk involved that the King's Arm is much worse than Les Epinards. On the other hand, it could also be much better. What should you do?

It all depends on how much information you have about either restaurant, and how many more days you'll stay in town. If this is your last day, then it's probably a better idea to go to Les Epinards, unless you are expecting King's Arm to be significantly better. However, if you are going to stay there longer, trying out King's Arm is a good bet. If you are lucky, you will be getting much better food for the remaining time, while otherwise you will have missed only one good meal out of many, making the potential risk quite small.

Thus, one must decide whether to *exploit* knowledge about the world to gain a *known* reward, or to *explore* the world to *learn* something new. This will potentially give you less reward immediately, but the knowledge itself can usually be put to use in the future.

This exploration-exploitation trade-off only arises when data collection is *interactive*. If we are simply given a set of data which is to be used to decide upon a course or action, but our decision does not affect the data we shall collect in the future, then things are much simpler. However, a lot of real-world human decision making as well as modern applications in data science involve such trade-offs. Decision theory offers precise mathematical models and algorithms for such problems.

1.3 Decision Theory and Reinforcement Learning

Decision theory deals with the formalization and solution of decision problems. Given a number of alternatives, what would be the rational choice in a particular situation depending on one's goals and desires? In order to answer this question we need to develop a good concept of *rational behavior*. This will serve two purposes. Firstly, this can provide an explanation for the behavior of animals and humans. Secondly, it is useful for developing models and algorithms for automated decision making in complex tasks.

A particularly interesting problem in this setting is *reinforcement learning*. This problem arises when the environment is unknown, and the learner has to make decisions solely through interaction, which only gives limited *feedback*. Thus, the learning agent does not have access to detailed instructions on which task to perform, nor on how to do it. Instead, it performs *actions*, which affect the environment and obtains some observations (i.e., sensory input) and feedback, usually in the form of *rewards* which represent the agent's desires. The learning problem is then formulated as the problem of learning how to act to maximize total reward. In biological systems, reward is intrinsically hardwired to signals associated with basic needs. In artificial systems, we can choose the reward signals so as to reinforce behaviour that achieves the designer's goals.

Reinforcement learning is a fundamental problem in artificial intelligence, since frequently we can tell robots, computers, or cars only what we would like them to achieve, but we do not know the best way to achieve it. We would like to simply give them a description of our goals and then let them explore the environment on their own to find a good solution. Since the world is (at least partially) unknown, the learner always has to deal with the exploration-exploitation trade-off.

Animals and humans also learn through imitation, exploration, and shaping their behavior according to reward signals to finally achieve their goals. In fact, it has been known since the 1990s that there is some connection between some reinforcement learning algorithms and mechanisms in the basal ganglia [1–3].

Decision theory is closely related to other fields, such as logic, statistics, game theory, and optimization. Those fields have slightly different underlying objectives, even though they may share the same formalism. In the field of *optimization*, we are not only interested in optimal planning in complex environments but also in how to make *robust* plans given some uncertainty about the environment. *Artificial intelligence* research is concerned with modelling the environments and developing algorithms that are able to learn by interaction with the environment or from demonstration by teachers. *Economics* and *game theory* deal with the problem of modeling the behavior of rational agents and with designing mechanisms (such as markets) that will give incentives to agents to behave in a certain way.

Beyond pure research, there are also many applications connected to decision theory. Commercial applications arise e.g. in advertising where one wishes to model the preferences and decision making of individuals. Decision problems also arise in *security*. There are many decision problems, especially in cryptographic and biometric authentication, but also in detecting and responding to intrusions in networked computer systems. Finally, in the natural sciences, especially in *biology and medicine*, decision theory offers a way to automatically design and run experiments and to optimally construct clinical trials.

Outline

1. Subjective probability and utility: the notion of subjective probability; eliciting priors; the concept of utility; expected utility
2. Decision problems: maximizing expected utility; maximin utility; regret
3. Estimation: estimation as conditioning; families of distributions closed under conditioning; conjugate priors; concentration inequalities; PAC and high-probability bounds; Markov Chain Monte Carlo; ABC estimation
4. Sequential sampling and optimal stopping: sequential sampling problems; the cost of sampling; optimal stopping; martingales.
5. Reinforcement learning I—Markov decision processes: belief and information state; bandit problems; Markov decision processes; backwards induction; value iteration; policy iteration; temporal differences; linear programming
6. Reinforcement learning II—Stochastic and approximation algorithms: Sarsa; Q-learning; stochastic value iteration; TD(λ)
7. Reinforcement learning III—Function approximation: features and the curse of dimensionality; approximate value iteration; approximate policy iteration; policy gradient
8. Reinforcement learning IV—Bayesian reinforcement learning: bounds on utility; Thompson sampling; stochastic branch and bound; sparse sampling; partially observable MDPs
9. Reinforcement learning V—Distribution-free reinforcement learning: stochastic and metric bandits; UCRL; bounds for Thompson sampling

References

1. Yin, H.H., Knowlton, B.J.: The role of the basal ganglia in habit formation. Nat. Rev. Neurosci. **7**(6), 464 (2006)
2. Barto, A.G.: Adaptive Critics and the Basal Ganglia, pp. 215–232. MIT Press (1995)
3. Schultz, W., Dayan, P., Read Montague, P.: A neural substrate of prediction and reward. Science **275**(5306), 1593–1599 (1997)

Chapter 2
Subjective Probability and Utility

2.1 Subjective Probability

In order to make decisions, we need to be able to make predictions about the possible outcomes of each decision. Usually, we have *uncertainty* about what those outcomes are. This can be due to *stochasticity*, which is frequently used to model games of chance and inherently unpredictable physical phenomena. It can also be due to *partial information*, a characteristic of many natural problems. For example, it might be hard to know at any single moment how much change you have in your wallet, whether you will be able to catch the next bus, or to remember where you left your keys.

In either case, this uncertainty can be expressed as a *subjective belief*. This does not have to correspond to reality. For example, some people believe, quite inaccurately, that if a coin comes up tails for a long time, it is quite likely to come up heads very soon. Or, you might quite happily believe your keys are in your pocket, only to realise that you left them at home as soon you arrive at the office.

In this book, we assume the view that subjective beliefs can be modelled as *probabilities*. This allows us to treat uncertainty due to stochasticity and due to partial information in a unified framework. In doing so, we have to define for each problem a space of possible outcomes Ω and specify an appropriate probability distribution.

2.1.1 Relative Likelihood

Let us start with the simple example of guessing whether a tossed coin will come up head or tails. In this case the sample space Ω would correspond to every possible way the coin can land. Since we are only interested in predicting which face will be up, let $A \subset \Omega$ be all those cases where the coin comes up heads, and $B \subset \Omega$ be the set of tosses where it comes up tails. Here $A \cap B = \emptyset$, but there may be some other events such as the coin becoming lost, so it does not necessarily hold that

C. Dimitrakakis and R. Ortner, *Decision Making Under Uncertainty and Reinforcement Learning*, Intelligent Systems Reference Library 223,
https://doi.org/10.1007/978-3-031-07614-5_2

$A \cup B = \Omega$. Nevertheless, we only care about whether A is more likely than B. As said, this likelihood may be based only on subjective beliefs. We can express that via the concept of relative likelihood:

The relative likelihood of two events A and B

- If A is *more* likely than B, then we write $A \succ B$, or equivalently $B \prec A$.
- If A is *as likely* as B, then we write $A \eqsim B$.

We also use $\succsim$ and $\precsim$ for *at least as likely as* and for *no more likely than*.

Let us now speak more generally about the case where we have defined an appropriate σ-field $\mathcal{F}$ on Ω. Then each element $A_i \in \mathcal{F}$ will be a subset of Ω. We now wish to define relative likelihood relations for the elements $A_i \in \mathcal{F}$.[1]

As we would like to use the language of probability to talk about likelihoods, we need to define a probability measure that agrees with our given relations. A probability measure $P : \mathcal{F} \to [0, 1]$ is said to *agree* with a relation $A \precsim B$, if it has the property that $P(A) \leq P(B)$ if and only if $A \precsim B$, for all $A, B \in \mathcal{F}$. In general, there are many possible measures that can agree with a given relation, cf. Example 2.1 below. However, it could also be that a given relational structure is incompatible with any possible probability measure. We also consider the question under which assumptions a likelihood relation corresponds to a unique probability measure.

2.1.2 Subjective Probability Assumptions

We would like our beliefs to satisfy some intuitive properties about what statements we can make concerning the relative likelihood of events. As we will see, these assumptions are also necessary to guarantee the existence of a corresponding probability measure. First of all, it must always be possible to say whether one event is more likely than the other, i.e., our beliefs must be complete. Consequently, we are not allowed to claim ignorance.

Assumption 2.1.1 (*SP1*) For any pair of events $A, B \in \mathcal{F}$, one has either $A \succ B$, $A \prec B$, or $A \eqsim B$.

Another important assumption is a principle of consistency: Informally, if we believe that every possible event A_i that leads to A is less likely than a unique corresponding event B_i that leads to an outcome B, then we should always conclude that A is less likely than B.

[1] More formally, we can define three classes: $C_{\succ}, C_{\prec}, C_{\eqsim} \subset \mathcal{F}^2$ such that a pair $(A_i, A_j) \in C_R$ if an only if it satisfies the relation $A_i R A_j$, where $R \in \{\succ, \prec, \eqsim\}$. These three classes form a partition of $\mathcal{F}^2$ under the subjective probability assumptions we will introduce in the next section.

Assumption 2.1.2 (*SP2*) Let $A = A_1 \cup A_2$, $B = B_1 \cup B_2$ with $A_1 \cap A_2 = B_1 \cap B_2 = \emptyset$. If $A_i \precsim B_i$ for $i = 1, 2$ then $A \precsim B$.

We also require the simple technical assumption that any event $A \in \mathcal{F}$ is at least as likely as the empty event $\emptyset$, which never happens.

Assumption 2.1.3 (*SP3*) For all A it holds that $\emptyset \precsim A$. Further, $\emptyset \prec \Omega$.

As it turns out, these assumptions are sufficient for proving the following theorems [1]. The first theorem tells us that our belief must be consistent with respect to transitivity.

Theorem 2.1.4 (Transitivity) *Under Assumptions 2.1.1, 2.1.2, and 2.1.3, for all events A, B, C: If $A \precsim B$ and $B \precsim C$, then $A \precsim C$.*

The second theorem says that if two events have a certain relation, then their negations have the converse relation.

Theorem 2.1.5 (Complement) *For any A, B: $A \precsim B$ iff $A^{\mathsf{C}} \succ B^{\mathsf{C}}$.*

Finally, note that if $A \subset B$, then it must be the case that whenever A happens, B must happen and hence B must be at least as likely as A.

Theorem 2.1.6 (Fundamental property of relative likelihoods) *If $A \subset B$ then $A \precsim B$. Furthermore, $\emptyset \precsim A \precsim \Omega$ for any event A.*

Since we are dealing with σ-fields, we need to introduce additional assumptions for infinite sequences of events. While these are not necessary if the field $\mathcal{F}$ is finite, it is good to include them for generality.

Assumption 2.1.7 (*SP4*) If $A_1 \supset A_2 \supset \cdots$ is a decreasing sequence of events in $\mathcal{F}$ and $B \in \mathcal{F}$ is such that $A_i \succsim B$ for all i, then $\bigcap_{i=1}^{\infty} A_i \succsim B$.

As a consequence, we obtain the following dual theorem:

Theorem 2.1.8 *If $A_1 \subset A_2 \subset \cdots$ is an increasing sequence of events in $\mathcal{F}$ and $B \in \mathcal{F}$ is such that $A_i \precsim B$ for all i, then $\bigcup_{i=1}^{\infty} A_i \precsim B$.*

We are now able to state a theorem for the unions of infinite sequences of disjoint events.

Theorem 2.1.9 *If $(A_i)_{i=1}^{\infty}$ and $(B_i)_{i=1}^{\infty}$ are infinite sequences of disjoint events in $\mathcal{F}$ such that $A_i \precsim B_i$ for all i, then $\bigcup_{i=1}^{\infty} A_i \precsim \bigcup_{i=1}^{\infty} B_i$.*

As said, our goal is to express likelihood via probability. Accordingly, we say that likelihood is induced by a probability measure P if for all A, B: $A \succ B$ iff $P(A) > P(B)$, and $A \eqsim B$ iff $P(A) = P(B)$. In this case the following theorem guarantees that the stipulated assumptions are always satisfied.

Theorem 2.1.10 *Let P be a probability measure over Ω. Then*

(i) $P(A) > P(B)$, $P(A) < P(B)$, or $P(A) = P(B)$ for all A, B.

(ii) Consider (possibly infinite) partitions $\{A_i\}_i$, $\{B_i\}_i$ of A, B, respectively. If $P(A_i) \leq P(B_i)$ for all i, then $P(A) \leq P(B)$.
(iii) For any A, $P(\emptyset) \leq P(A)$ and $P(\emptyset) < P(\Omega)$.

Proof Part (i) is trivial, as $P : \mathcal{F} \rightarrow [0, 1]$. Part (ii) holds due to $P(A) = P(\bigcup_i A_i) = \sum_i P(A_i) \leq \sum_i P(B_i) = P(B)$. Part (iii) follows from $P(\emptyset) = 0$, $P(A) \geq 0$, and $P(\Omega) = 1$. □

2.1.3 Assigning Unique Probabilities*

In many cases, and particularly when $\mathcal{F}$ is a finite field, there is a large number of probability distributions agreeing with our relative likelihoods. Choosing one specific probability over another does not seem easy. The following example underscores this ambiguity.

Example 2.1 Consider $\mathcal{F} = \{\emptyset, A, A^{\mathsf{C}}, \Omega\}$ for some A with $\emptyset \neq A \subseteq \Omega$ and assume $A \succ A^{\mathsf{C}}$. Consequently, $P(A) > 1/2$, which is however insufficient for assigning a specific value to $P(A)$.

In some cases we would like to assign unique probabilities to events in order to facilitate computations. Intuitively, we may only be able to tell whether some event A is more likely than some other event B. However, we can create a new, uniformly distributed random variable x on $[0, 1]$ and determine for each value $\alpha \in [0, 1]$ whether A more or less likely than the event $x > \alpha$. Since we need to compare both A and B with all such events, the distribution we obtain is unique. For the sake of completeness we start with the definition of the uniform distribution over the interval $[0, 1]$.

Definition 2.1.11 (*Uniform distribution*) Let $\lambda(A)$ denote the length of any interval $A \subseteq [0, 1]$. Then $x : \Omega \rightarrow [0, 1]$ has a uniform distribution on $[0, 1]$ if, for any subintervals A, B of $[0, 1]$,

$$(x \in A) \precsim (x \in B) \quad \text{iff} \quad \lambda(A) \leq \lambda(B),$$

where $(x \in A)$ denotes the event that $x(\omega) \in A$. Then $(x \in A) \precsim (x \in B)$ means that ω is such that $x \in A$ is not more likely than $x \in B$.

This means that *any* larger interval is more likely than *any* smaller interval. Now we shall connect the uniform distribution to the original sample space Ω by assuming that there is some function with uniform distribution.

Assumption 2.1.12 (*SP5*) It is possible to construct a random variable $x : \Omega \rightarrow [0, 1]$ with a uniform distribution in $[0, 1]$.

Constructing the Probability Distribution

We can now use the uniform distribution to create a unique probability measure that agrees with our likelihood relation. First, we have to map each event in Ω to an equivalent event in $[0, 1]$, which can be done according to the following theorem.

Theorem 2.1.13 (Equivalent event) *For any event $A \in \mathcal{F}$, there exists some $\alpha \in [0, 1]$ such that $A \eqsim (x \in [0, \alpha])$.*

This means that we can now define the probability of an event A by matching it to a specific equivalent event on $[0, 1]$.

Definition 2.1.14 (*Probability of A*) Given any event A, define $P(A)$ to be the α with $A \eqsim (x \in [0, \alpha])$.

Hence

$$A \eqsim (x \in [0, P(A)]),$$

which is sufficient to show the following theorem.

Theorem 2.1.15 (Relative likelihood and probability) *If assumptions SP1–SP5 are satisfied, then the probability measure P defined in Definition 2.1.14 is unique. Furthermore, for any two events A, B it holds that $A \precsim B$ iff $P(A) \leq P(B)$.*

2.1.4 Conditional Likelihoods

So far we have only considered the problem of forming opinions about which events are more likely *a priori*. However, we also need to have a way to incorporate evidence which may adjust our opinions. For example, while we ordinarily may think that $A \precsim B$, we may have additional information D, given which we think the opposite is true. We can formalise this through the notion of conditional likelihoods.

Example 2.2 Say that A is the event that it rains in Gothenburg, Sweden tomorrow. We know that Gothenburg is quite rainy due to its oceanic climate, so we set $A \succsim A^{\complement}$. Now, let us try and incorporate some additional information. Let D denote the fact that good weather is forecast, so that given D good weather will appear more probable than rain, formally $(A \mid D) \precsim (A^{\complement} \mid D)$.

Conditional Likelihoods

Define $(A \mid D) \precsim (B \mid D)$ to mean that B is at least as likely as A when it is known that D holds.

Assumption 2.1.16 (*CP*) For any events A, B, D,

$$(A \mid D) \precsim (B \mid D) \quad \text{iff} \quad A \cap D \precsim B \cap D.$$

Theorem 2.1.17 *If a likelihood relation $\precsim$ satisfies assumptions SP1–SP5 as well as CP, then there exists a probability measure P such that: For any A, B, D such that $P(D) > 0$,*

$$(A \mid D) \precsim (B \mid D) \quad \textit{iff} \quad P(A \mid D) \leq P(B \mid D).$$

It turns out that there are very few ways that a conditional probability definition can satisfy all of our assumptions. The usual definition is the following.

Definition 2.1.18 (*Conditional probability*)

$$P(A \mid D) \triangleq \frac{P(A \cap D)}{P(D)}.$$

This definition effectively answers the question of how much evidence for A we have, now that we have observed D. This is expressed as the ratio between the combined event $A \cap D$, also known as the joint probability of A and D, and the marginal probability of D itself. The intuition behind the definition becomes clearer once we rewrite it as $P(A \cap D) = P(A \mid D) P(D)$. Then conditional probability is effectively used as a way to factorise joint probabilities.

2.1.5 *Probability Elicitation*

Probability elicitation is the problem of quantifying the subjective probabilities of a particular individual. One of the simplest, and most direct, methods, is to simply ask. However, because we cannot simply ask somebody to completely specify a probability distribution, we can ask for this distribution iteratively as indicated by the procedure outlined in the following example.

Example 2.3 (*Temperature prediction*) Let τ be the temperature tomorrow at noon in Gothenburg. What are your estimates?

Eliciting the prior/forming the subjective probability measure P

- Select temperature x_0 s.t. $(\tau \leq x_0) \eqsim (\tau > x_0)$.
- Select temperature x_1 s.t. $(\tau \leq x_1 \mid \tau \leq x_0) \eqsim (\tau > x_1 \mid \tau \leq x_0)$.

By repeating this procedure recursively we are able to slowly build the complete distribution, quantile by quantile.

Note that, necessarily, $P(\tau \leq x_0) = P(\tau > x_0) =: p_0$. Since $P(\tau \leq x_0) + P(\tau > x_0) = P(\tau \leq x_0 \cup \tau > x_0) = P(\tau \in \mathbb{R}) = 1$, it follows that $p_0 = \frac{1}{2}$. Similarly, $P(\tau \leq x_1 \mid \tau \leq x_0) = P(\tau > x_1 \mid \tau \leq x_0) = \frac{1}{4}$.

Exercise 2.1.19 Propose another way to arrive at a prior probability distribution. For example, define a procedure for eliciting a single probability distribution from a group of people without any interaction between the participants.

2.2 Updating Beliefs: Bayes' Theorem

Although we always start with a particular belief, this belief must be adjusted when we receive new evidence. In probabilistic inference, the updated beliefs are simply the probability of future events conditioned on observed events. This idea is captured neatly by Bayes' theorem, which links the prior probability $P(A_i)$ of events to their posterior probability $P(A_i \mid B)$ given some event B and the probability $P(B \mid A_i)$ of observing the evidence B given events A_i.

Theorem 2.2.1 (Bayes' theorem) *Let $A_1, A_2, \ldots$ be a (possibly infinite) sequence of disjoint events such that $\bigcup_{i=1}^{n} A_i = \Omega$ and $P(A_i) > 0$ for all i. Let B be another event with $P(B) > 0$. Then*

$$P(A_i \mid B) = \frac{P(B \mid A_i)\, P(A_i)}{\sum_{j=1}^{n} P(B \mid A_j)\, P(A_j)}.$$

Proof By definition, $P(A_i \mid B) = P(A_i \cap B)/P(B)$ and $P(A_i \cap B) = P(B \mid A_i)\, P(A_i)$, so

$$P(A_i \mid B) = \frac{P(B \mid A_i)\, P(A_i)}{P(B)}. \tag{2.1}$$

As $\bigcup_{j=1}^{n} A_j = \Omega$, we have $B = \bigcup_{j=1}^{n}(B \cap A_j)$. Since the A_j are disjoint, so are the $B \cap A_j$. As P is a probability, the union property gives

$$P(B) = P\left(\bigcup_{j=1}^{n}(B \cap A_j)\right) = \sum_{j=1}^{n} P(B \cap A_j) = \sum_{j=1}^{n} P(B \mid A_j)\, P(A_j),$$

which plugged into (2.1) completes the proof. □

A Simple Exercise in Updating Beliefs

Example 2.4 (*The weather forecast*) Form a subjective probability for the probability that it rains.

A_1 : Rain.
A_2 : No rain.

First, choose $P(A_1)$ and set $P(A_2) = 1 - P(A_1)$. Now assume that there is a weather forecasting station that predicts *no rain* for tomorrow. However, you know the following facts about the station: On the days when it rains, half of the time the station had predicted it was *not* going to rain. On days when it doesn't rain, the station had said *no rain* 9 times out of 10.

(*Solution*) Let B denote the event that the station predicts no rain. According to our information, the probability that there is rain when the prediction said no rain is $P(B \mid A_1) = \frac{1}{2}$. On the other hand, $P(B \mid A_2) = 0.9$. Combining these with Bayes' theorem, we obtain

$$P(A_1 \mid B) = \frac{P(B \mid A_1)\, P(A_1)}{P(B \mid A_1)\, P(A_1) + P(B \mid A_2)\, [1 - P(A_1)]} = \frac{\frac{1}{2} P(A_1)}{0.9 - 0.4\, P(A_1)}.$$

□

2.3 Utility Theory

While probability can be used to describe how likely an event is, utility can be used to describe how desirable it is. More concretely, our subjective probabilities are numerical representations of our beliefs and the information available to us. They can be taken to represent our "internal model" of the world. By analogy, our utilities are numerical representations of our tastes and preferences. That is, even if the consequences of our actions are not directly known to us, we assume that we act to maximise our utility, in some sense.

2.3.1 *Rewards and Preferences*

Rewards

Consider that we have to choose a *reward* r from a set $\mathcal{R}$ of possible rewards. While the elements of $\mathcal{R}$ may be arbitrary, we shall in general find that we prefer some rewards to others. In fact, some elements of $\mathcal{R}$ may not even be desirable. As an example, $\mathcal{R}$ might be a set of tickets to different musical events, or a set of financial commodities.

Preferences

Example 2.5 (*Musical event tickets*) We have a set of tickets $\mathcal{R}$, and we must choose the ticket $r \in \mathcal{R}$ we prefer best. Here are two possible scenarios:

- $\mathcal{R}$ is a set of tickets to different music events at the same time, at equally good halls with equally good seats and the same price. Here preferences simply coincide with the preferences for a certain type of music or an artist.
- If $\mathcal{R}$ is a set of tickets to different events at different times, at different quality halls with different quality seats and different prices, preferences may depend on all the factors.

Example 2.6 (*Route selection*) We have a set of alternative routes and must pick one. If $\mathcal{R}$ contains two routes of the same quality, one short and one long, we will probably prefer the shorter one. If the longer route is more scenic our preferences may be different.

Preferences Among Rewards

We will treat preferences in a similar manner as we have treated likelihoods. That is, we will define a linear ordering among possible rewards.

Let $a, b \in \mathcal{R}$ be two rewards. When we *prefer a* to b, we write $a \succ^* b$. Conversely, when we like a *less* than b we write $a \prec^* b$. If we like a *as much* as b, we write $a \eqsim^* b$. We also use $a \succsim^* b$ and $a \precsim^* b$ when we *like a at least as much as b* and we *don't like a any more than b*, respectively.

We make the following assumptions about the preference relations.

Assumption 2.3.1 (i) For any $a, b \in \mathcal{R}$, one of the following holds: $a \succ^* b, a \prec^* b$, or $a \eqsim^* b$.
(ii) For any $a, b, c \in \mathcal{R}$, if $a \precsim^* b$ and $b \precsim^* c$, then $a \precsim^* c$.

The first assumption means that we must always be able to decide between any two rewards. It may seem that it does not always hold in practice, since humans are frequently indecisive. However, without the second assumption, it is still possible to create preference relations that are cyclic.

Example 2.7 (*Counter-example for transitive preferences*) Consider vector rewards in $\mathcal{R} = \mathbb{R}^2$, with $r_i = (a_i, b_i)$, and fix some $\epsilon > 0$. Our preference relation satisfies:

- $r_i \succ^* r_j$ if $b_i \geq b_j + \epsilon$.
- $r_i \succ^* r_j$ if $a_i > a_j$ and $|b_i - b_j| < \epsilon$.

This may correspond for example to an employer deciding to hire one out of several candidates depending on their experience (a_i) and their school grades (b_i). Since grades are not very reliable, if two people have similar grades, then we prefer the one with more experience. However, that may lead to a cycle. Consider a sequence of candidates $i = 1, \ldots, n$, such that $b_i = b_{i+1} + \delta$, with $\delta < \epsilon$ and $a_i > a_{i+1}$. Then clearly, we must always prefer r_i to r_{i+1}. However, if $\delta(n-1) > \epsilon$, we will prefer r_n to r_1.

2.3.2 *Preferences Among Distributions*

When We Cannot Select Rewards Directly

In most problems, we cannot choose the rewards directly. Rather, we must make some decision, and then obtain a reward depending on this decision. Since we may be uncertain about the outcome of a decision, we can specify our uncertainty regarding the rewards obtained by a decision in terms of a probability distribution.

Example 2.8 (*Route selection*) Assume that you have to pick between two routes P_1, P_2. Your preferences are such that shorter time routes are preferred over longer ones. For simplicity, let $\mathcal{R} = \{10, 15, 30, 35\}$ be the possible times it might take to reach your destination. Route P_1 takes 10 min when the road is clear, but 30 min when the traffic is heavy. The probability of heavy traffic on P_1 is 0.5. On the other hand, route P_2 takes 15 min when the road is clear, but 35 min when the traffic is heavy. The probability of heavy traffic on P_2 is 0.2.

Preferences Among Probability Distributions

As seen in the previous example, we frequently have to define preferences between probability distributions, rather than over rewards. To represent our preferences, we can use the same notation as before. Let P_1, P_2 be two distributions on $(\mathcal{R}, \mathcal{F}_{\mathcal{R}})$. If we *prefer* P_1 to P_2, we write $P_1 \succ^* P_2$. If we like P_1 *less* than P_2, write $P_1 \prec^* P_2$. If we like P_1 *as much* as P_2, we write $P_1 \eqsim^* P_2$. Finally, as before we also use $\succsim^*$ and $\precsim^*$ to denote negations of the strict preference relations $\succ^*$ and $\prec^*$.

What would be a good principle for choosing between the two routes in Example 2.8? Clearly route P_1 gives both the lowest best-case time and the lowest worst-case time. It thus appears as though both an extremely cautious person (who assumes the worst-case) and an extreme optimist (who assumes the best case) would say $P_1 \succ^* P_2$. However, the average time taken in P_2 is only 19 min versus 20 min for P_1. Thus, somebody who only took the average time into account would prefer P_2. In the following sections, we will develop one of the most fundamental methodologies for choices under uncertainty, based on the idea of utilities.

2.3.3 *Utility*

The concept of utility allows us to create a unifying framework, such that given a particular set of rewards and probability distributions on them, we can define preferences among distributions via their expected utility. The first step is to define utility as a way to define a preference relation among rewards.

Definition 2.3.2 (*Utility*) A utility function $U : \mathcal{R} \to \mathbb{R}$ is said to agree with the preference relation $\succsim^*$, if for all rewards $a, b \in \mathcal{R}$

$$a \succsim^* b \quad \text{iff} \quad U(a) \geq U(b).$$

The above definition is very similar to how we defined relative likelihood in terms of probability. For a given utility function, its expectation for a distribution over rewards is defined as follows.

Table 2.1 A simple gambling problem

r	$U(r)$	P	Q
Did not enter	0	1	0
Paid 1 CU and lost	-1	0	0.99
Paid 1 CU and won 10	9	0	0.01

Definition 2.3.3 (*Expected utility*) Given a utility function U, the expected utility of a distribution P on $\mathcal{R}$ is

$$\mathbb{E}_P(U) = \int_{\mathcal{R}} U(r)\, \mathrm{d}P(r).$$

We make the assumption that the utility function is such that the expected utility remains consistent with the preference relations between all probability distributions we are choosing between.

Assumption 2.3.4 (*The expected utility hypothesis*) Given a preference relation $\succsim^*$ over $\mathcal{R}$ and a corresponding utility function U, the utility of any probability measure P on $\mathcal{R}$ is equal to the expected utility of the reward under P. Consequently,

$$P \succsim^* Q \quad \text{iff} \quad \mathbb{E}_P(U) \geq \mathbb{E}_Q(U). \tag{2.2}$$

Example 2.9 Consider the following decision problem. You have the option of entering a lottery for 1 currency unit (CU). The prize is 10 CU and the probability of winning is 0.01. This can be formalised by making it a choice between two probability distributions: P, where you do not enter the lottery, and Q, which represents entering the lottery.

Calculating the expected utility obviously gives 0 for not entering and $\mathbb{E}(U \mid Q) = \sum_r U(r)\, Q(r) = -0.9$ utility of entering the lottery, cf. Table 2.1.

Monetary Rewards

Frequently, rewards come in the form of money. In general, it is assumed that people prefer to have more money than less money. However, while the utility of monetary rewards is generally assumed to be increasing, it is not necessarily linear. For example, 1,000 Euros are probably worth more to somebody with only 100 Euros in the bank than to somebody with 100,000 Euros. Hence, it seems reasonable to assume that the utility of money is concave.

The following examples show the consequences of the expected utility hypothesis.

Example 2.10 Choose between the following two gambles:

1. The reward is 500,000 with certainty.
2. The reward is 2,500,000 with probability 0.10. It is 500,000 with probability 0.89, and 0 with probability 0.01.

Example 2.11 Choose between the following two gambles:

1. The reward is 500,000 with probability 0.11, or 0 with probability 0.89.
2. The reward is 2,500,000 with probability 0.1, or 0 with probability 0.9.

Exercise 2.3.5 Show that under the expected utility hypothesis, if gamble 1 is preferred in the Example 2.10, gamble 1 must also be preferred in the Example 2.11 for any utility function.

In practice, you may find that your preferences are not aligned with what this exercise suggests. This implies that either your decisions do not conform to the expected utility hypothesis, or that you are not internalising the given probabilities. We will explore this further in following example.

The St. Petersburg Paradox

The following simple example illustrates the fact that, internally, most humans do not behave in ways that are compatible with linear utility for money.

Example 2.12 (*The St. Petersburg Paradox, Bernoulli 1713*) A coin is tossed repeatedly until the coin comes up heads. The player obtained 2^n currency units, where $n \in \{1, 2, \dots\}$ is the number of times the coin was thrown. The coin is assumed to be *fair*, meaning that the probability of heads is always $1/2$.

How many currency units k would you be willing to pay to play this game once?

As the probability to stop at round n is 2^{-n}, the expected monetary gain of the game is

$$\sum_{n=1}^{\infty} 2^n 2^{-n} = \infty.$$

Were your utility function linear, you would be willing to pay any finite amount k to play, as the expected utility for playing the game for any finite k is

$$\sum_{n=1}^{\infty} U(2^n - k)\, 2^{-n} = \infty.$$

It would be safe to assume that very few readers would be prepared to pay an arbitrarily high amount to play this game. One way to explain this is that the utility function is not linear. An alternative would be to assume a logarithmic utility function. For example, if we also assume that the player has an initial capital C from which k has to be paid, the utility function would satisfy

$$\mathbb{E}\, U = \sum_{n=1}^{\infty} \ln(C + 2^n - k)\, 2^{-n}.$$

Then for $C = 10,000$ the maximum bet would be 14. For $C = 100$ it would be 6, while for $C = 10$ it is just 4.

There is another reason why one may not pay an arbitrary amount to play this game. The player may not fully internalise the fact (or rather, the promise) that the coin is unbiased. Another explanation would be that it is not really believed that the bank can pay an unbounded amount of money. Indeed, if the bank can only pay amounts up to N, in the linear expected utility scenario, for a coin with probability p of coming heads, we have

$$\sum_{n=1}^{N} 2^n p^{n-1}(1-p) = 2(1-p)\frac{1-(2p)^N}{1-2p}.$$

For large N and $p = 0.45$, it turns out that you should only expect a payoff of about 10 currency units. Similarly, for a fair coin and $N = 1024$ you should only pay around 10 units as well. These are possible subjective beliefs that an individual might have that would influence its behaviour when dealing with a formally specified decision problem.

2.3.4 *Measuring Utility**

Since we cannot even rely on linear utility for money, we need to ask ourselves how we can measure the utility of different rewards. There are a number of ways, including trying to infer it from the actions of people. The simplest approach is to simply ask them to make even money bets. No matter what approach we use, however, we need to make some assumptions about the utility structure. This includes whether or not we should accept that the expected utility hypothesis holds for the observed human behaviour.

Experimental Measurement of Utility

Example 2.13 Let $\langle a, b \rangle$ denote a lottery ticket that yields a or b CU with equal probability. Consider the following sequence:

1. Find x_1 such that receiving x_1 CU with certainty is equivalent to receiving $\langle a, b \rangle$.
2. Find x_2 such that receiving x_2 CU with certainty is equivalent to receiving $\langle a, x_1 \rangle$.
3. Find x_3 such that receiving x_3 CU with certainty is equivalent to receiving $\langle x_1, b \rangle$.
4. Find x_4 such that receiving x_4 CU with certainty is equivalent to receiving $\langle x_2, x_3 \rangle$.

The above example algorithm allows us to measure the utility of money under the assumption that the expected utility hypothesis holds. Note that if $x_1 \neq x_4$, then the

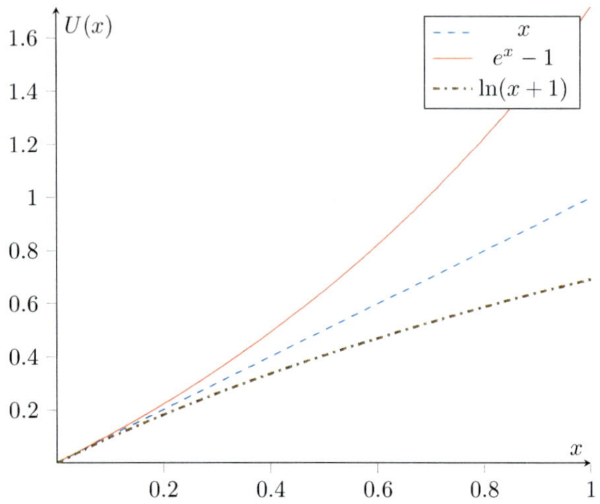

Fig. 2.1 Linear, convex and concave functions

expected utility hypothesis is violated, since it implies $U(x_1) = U(x_4) = \frac{1}{2}(U(a) + U(b))$.

2.3.5 *Convex and Concave Utility Functions*

As previously mentioned, utility functions of monetary rewards seem to be concave. In general, we would say that a concave utility function implies risk aversion and a convex one risk taking. Intuitively, a risk averse person prefers a fixed amount of money to a random amount of money with the same expected value. A risk taker prefers to gamble. Let's start with the definition of a convex function, where in the following we assume that Ω is a convex subset of $\mathbb{R}^n$, that is, Ω contains with any two points x, y also the line segment between x and y.

Definition 2.3.6 A function $g : \Omega \to \mathbb{R}$ is convex on $A \subset \Omega$ if for any points $x, y \in A$ and any $\alpha \in [0, 1]$

$$\alpha g(x) + (1 - \alpha)g(y) \geq g(\alpha x + (1 - \alpha)y).$$

An important property of convex functions is that they are bounded from above by linear segments connecting their points. This property is formally given below.

Theorem 2.3.7 (Jensen's inequality) *Let g be a convex function on Ω and P be a measure with $P(\Omega) = 1$. Then for any $x \in \Omega$ such that $\mathbb{E}(x)$ and $\mathbb{E}[g(x)]$ exist, it holds that*

$$\mathbb{E}[g(x)] \geq g[\mathbb{E}(x)].$$

Example 2.14 If the utility function is convex, then we would prefer obtaining a random reward x rather than a fixed reward $y = \mathbb{E}(x)$. Thus, a convex utility function implies risk-taking. This is illustrated by Fig. 2.1, which shows a linear function, x, a convex function, $e^x - 1$, and a concave function, $\ln(x + 1)$.

Definition 2.3.8 A function g is concave on $A \subset \Omega$ if for any points $x, y \in A$ and any $\alpha \in [0, 1]$

$$\alpha g(x) + (1 - \alpha)g(y) \leq g(\alpha x + (1 - \alpha)y).$$

For concave functions, the inverse of Jensen's inequality holds (i.e., with $\geq$ replaced with $\leq$). If the utility function is concave, then we choose a gamble giving a fixed reward $\mathbb{E}[x]$ rather than one giving a random reward x. Consequently, a concave utility function implies risk aversion. The act of buying insurance can be related to the concavity of our utility function. Consider the following example, where we assume individuals are risk-averse, but insurance companies are risk-neutral.

Example 2.15 (*Insurance*) Let d be the insurance cost, h our insurance cover, ϵ the probability of needing the cover, and U an increasing utility function (for monetary values). Then we are going to buy insurance if the utility of losing d with certainty is greater than the utility of losing $-h$ with probability ϵ.

$$U(-d) > \epsilon\, U(-h) + (1 - \epsilon)\, U(0). \tag{2.3}$$

The insurance company has a linear utility and fixes the premium d high enough so that

$$d > \epsilon h. \tag{2.4}$$

Consequently, we see from (2.4) that $U(-\epsilon h) \geq U(-d)$, as U is an increasing function. From (2.3) we obtain $U(-\epsilon h) > \epsilon U(-h) + (1 - \epsilon)\, U(0)$. Now the $-\epsilon h$ term is the utility of our expected monetary loss, while the right hand side is our expected utility. Consequently if the inequality holds, our utility function is (at least locally) concave.

2.4 Exercises

Exercise 2.4.1 Preferences are transitive if they are induced by a utility function $U : \mathcal{R} \to \mathbb{R}$ such that $a \succ^* b$ iff $U(a) > U(b)$. Give an example of a utility function, not necessarily mapping the rewards $\mathcal{R}$ to $\mathbb{R}$, and a binary relation $>$ such that transitivity can be violated. Back your example with a thought experiment.

Exercise 2.4.2 Assuming that U is increasing and absolutely continuous, consider the following experiment:

1. You specify an amount a, then observe random value Y.
2. If $Y \geq a$, you receive Y currency units.
3. If $Y < a$, you receive a random amount X with known distribution (independent of Y).

Show that we should choose a such that $U(a) = \mathbb{E}[U(X)]$.

Exercise 2.4.3 (*Usefulness of probability and utility*)

1. Would it be useful to separate randomness from uncertainty? What would be desirable properties of an alternative concept to probability?
2. Give an example of how the expected utility assumption might be violated.

Exercise 2.4.4 Consider two urns, each containing red and blue balls. The first urn contains an equal number of red and blue balls. The second urn contains a *randomly* chosen proportion X of red balls, i.e., the probability of drawing a red ball from that urn is X.

1. Suppose that you were to select an urn, and then choose a random ball from that urn. If the ball is red, you win 1 CU, otherwise nothing. Show that if your utility function is increasing with monetary gain, you should prefer the first urn iff $\mathbb{E}(X) < \frac{1}{2}$.
2. Suppose that you were to select an urn, and then choose n random balls from that urn and that urn only. Each time you draw a red ball, you gain 1 CU. After you draw a ball, you put it back in the urn. Assume that the utility U is strictly concave and suppose that $\mathbb{E}(X) = \frac{1}{2}$. Show that you should always select balls from the first urn.

Hint: Show that for the second urn, $\mathbb{E}(U \mid x)$ *is concave for* $0 \leq x \leq 1$ *(this can be done by showing* $\frac{d^2}{dx^2}\mathbb{E}(U \mid x) < 0$*).* In fact,

$$\frac{d^2}{dx^2}\mathbb{E}(U \mid x) = n(n-1)\sum_{k=0}^{n-2}[U(k) - 2U(k+1) + U(k+2)]\binom{n-2}{k}x^k(1-x)^{n-2-k}.$$

Then apply Jensen's inequality.

Exercise 2.4.5 (*Defining likelihood relations via Probability measures*) Show that a probability measure P on $(\Omega, \mathcal{F})$ satisfies the following:

1. For any events $A, B \in \mathcal{F}$ either $P(A) > P(B)$, $P(B) > P(A)$ or $P(A) = P(B)$.
2. If A_i, B_i are partitions of A, B such that for all $P(A_i) \leq P(B_i)$ for all i, then $P(A) \leq P(B)$.
3. For any event A, $P(\emptyset) \leq P(A)$ and $P(\emptyset) < P(\Omega)$.

Exercise 2.4.6 (*Definition of conditional probability*) Recall that $P(A \mid B) \triangleq \frac{P(A \cap B)}{P(B)}$ is only a definition. Think of a a plausible alternative for a definition of conditional probability. Note that the conditional probability $P(A \mid B)$ is just a new probability measure that shall satisfy the following basic properties of a probability measure: (a) null probability: $P(\emptyset \mid B) = 0$, (b) total probability: $P(\Omega \mid B) = 1$, (c) union of disjoint subsets: $P(A_1 \cup A_2 \mid B) = P(A_1 \mid B) + P(A_2 \mid B)$, (d) conditional probability: $P(A \mid D) \leq P(B \mid D)$ if and only if $P(A \cap D) \leq P(B \cap D)$.

Exercise 2.4.7 (*Alternatives to the expected utility hypothesis*) The expected utility hypothesis states that we prefer decision P over Q if and only if our expected utility under the distribution P is larger than that under Q, i.e., $\mathbb{E}_P(U) \geq \mathbb{E}_Q(U)$. Under what conditions do you think this is a reasonable hypothesis? Can you come up with a different rule for making decisions under uncertainty? Would it still satisfy the total order and transitivity properties of preference relations? In other words, could you still unambiguously say for any P, Q whether you prefer P to Q? If you had three choices, P, Q, W, and you preferred P to Q and Q to W, would you always prefer P to W?

Exercise 2.4.8 (*Rational Arthur-Merlin games*) You are Arthur, and you wish to pay Merlin to do a very difficult computation for you. More specifically, you perform a query $q \in Q$ and obtain an answer $r \in R$ from Merlin. After he gives you the answer, you give Merlin a random amount of money m, depending on r, q. In particular, it is assumed that there exists a unique correct answer $r^* = f(q)$ and $\mathbb{E}(m \mid r, q) = \sum_m m P(m \mid r, q)$ is maximized by r^*, i.e., for any $r \neq r^*$

$$\mathbb{E}(m \mid r^*, q) > \mathbb{E}(m \mid r, q).$$

Assume that Merlin knows P and the function f. Is this sufficient to incentivize Merlin to respond with the correct answer? If not, what other assumptions or knowledge are required?

Exercise 2.4.9 Assume that you need to travel over the weekend. You wish to decide whether to take the train or the car. Assume that the train and the car trip cost exactly the same amount of money. The train trip takes 2 h. If it does not rain, then the car trip takes 1.5 h. However, if it rains the road becomes both more slippery and more crowded and so the average trip time is 2.5 h. Assume that your utility function is equal to the negative amount of time spent travelling: $U(t) = -t$.

1. Let it be Friday. What is the expected utility of taking the car on Sunday? What is the expected utility of taking the train on Sunday? What is the Bayes-optimal decision, assuming you will travel on Sunday?
2. Consider two stations H_1 and H_2 that predict rain on Saturday and Sunday with probabilities 0.4, 0.6 (H_1) and 0.9, 0.1 (H_2), respectively. Let it be a rainy Saturday, which we denote by event A. What is your posterior probability over the two weather stations, given that it has rained, i.e., $P(H_i \mid A)$? What is the new marginal probability of rain on Sunday, i.e., $P(B \mid A)$? What is now the expected utility of taking the car versus taking the train on Sunday? What is the Bayes-optimal decision?

Exercise 2.4.10 Consider the previous example with a nonlinear utility function.

1. One example is $U(t) = 1/t$, which is a *convex* utility function. How would you interpret the utility in that case? Without performing the calculations, can you tell in advance whether your optimal decision can change? Verify your answer by calculating the expected utility of the two possible choices.
2. How would you model a problem where the objective involves arriving in time for a particular appointment?

Reference

1. DeGroot, M.H.: Optimal Statistical Decisions. Wiley (1970)

Chapter 3
Decision Problems

3.1 Introduction

In this chapter we describe how to formalize statistical decision problems. These involve making decisions whose utility depends on an unknown *state of the world*. In this setting, it is common to assume that the state of the world is a fundamental property that is not influenced by our decisions. However, we can calculate a probability distribution for the state of the world, using a prior belief and some data, where the data we obtain may depend on our decisions.

A classical application of this framework is parameter estimation. Therein, we stipulate the existence of a parametrized *law of nature*, and we wish to choose a best-guess set of parameters for the law through measurements and some prior information. An example would be determining the gravitational attraction constant from observations of planetary movements. These measurements are always obtained through experiments, the automatic design of which will be covered in later chapters.

The decisions we make will necessarily depend on both our prior belief and the data we obtain. In the last section of this chapter will examine how sensitive our decisions are to the prior, and how we can choose it so that our decisions are robust.

3.2 Rewards that Depend on the Outcome of an Experiment

Consider the problem of choosing one of two different types of tickets in a raffle. Each type of ticket gives you the chance to win a different prize. The first is a bicycle and the second is a tea set. The winner ticket for each prize is drawn uniformly from the respective sold tickets. Thus, the raffle guarantees that somebody will win either price. If most people opt for the bicycle, your chance of actually winning it by buying a single ticket is much smaller. However, if you prefer winning a bicycle to winning

C. Dimitrakakis and R. Ortner, *Decision Making Under Uncertainty and Reinforcement Learning*, Intelligent Systems Reference Library 223,
https://doi.org/10.1007/978-3-031-07614-5_3

the tea set, it is not clear what choice you should make in the raffle. The above is the quintessential example for problems where the reward that we obtain depends not only on our decisions, but also on the outcome of an *experiment*.

This problem can be viewed more generally for scenarios where the reward you receive depends not only on your own choice, but also on some other, unknown fact in the world. This may be something completely uncontrollable, and hence you only can make an informed guess.

More formally, given a set of possible actions $\mathcal{A}$, we must make a decision $a \in \mathcal{A}$ *before* knowing the outcome ω of an experiment with outcomes in Ω. After the experiment is performed, we obtain a *reward* $r \in \mathcal{R}$ which depends on both the outcome ω of the experiment and our decision. As discussed in the previous chapter, our preferences for some rewards over others are determined by a *utility* function $U : \mathcal{R} \to \mathbb{R}$, such that we prefer r to r' if and only if $U(r) \geq U(r')$. Now, however, we cannot choose rewards directly. Another example, which will be used throughout this section, is the following.

Example 3.1 (Taking the umbrella) We must decide whether to take an umbrella to work. Our reward depends on whether we get wet and the amount of objects that we carry. We would rather not get wet and not carry too many things, which can be made more precise by choosing an appropriate utility function. For example, we might put a value of -1 for carrying the umbrella and a value of -10 for getting wet. In this example, the only events of interest are whether it rains or not.

3.2.1 Formalisation of the Problem Setting

The elements we need to formulate the problem setting are a random variable, a decision variable, a reward function mapping the random and the decision variable to a reward, and a utility function that says how much we prefer each reward.

Assumption 3.2.1 (*Outcomes*) There exists a probability measure P on $(\Omega, \mathcal{F}_\Omega)$ such that the probability of the random outcome ω being in $A \in \mathcal{F}_\Omega$ is

$$\mathbb{P}(\omega \in A) = P(A).$$

The probability measure P is completely independent of any decision that we make.

Assumption 3.2.2 (*Utilities*) Given a set of rewards $\mathcal{R}$, our preferences satisfy Assumptions 2.1.1, 2.1.2, and 2.1.3, i.e., preferences are transitive, all rewards are comparable, and there exists a utility function U, measurable with respect to $\mathcal{F}_\mathcal{R}$ such that $U(r) \geq U(r')$ iff $r \succ^* r'$.

Since the random outcome ω does not depend on our decision a, we must find a way to connect the two. This can be formalized via a reward function, so that the reward that we obtain (whether we get wet or not) depends on both our decision (to take the umbrella) and the random outcome (whether it rains).

Definition 3.2.1 (*Reward function*) A reward function $\rho : \Omega \times \mathcal{A} \to \mathcal{R}$ defines the reward we obtain if we select $a \in \mathcal{A}$ and the experimental outcome is $\omega \in \Omega$:

$$r = \rho(\omega, a)$$

When we discussed the problem of choosing between distributions in Sect. 2.3.2, we had directly defined probability distributions on the set of rewards. We can now formulate our problem in that setting. First, we define a set of distributions $\{P_a \mid a \in \mathcal{A}\}$ on the reward space $(\mathcal{R}, \mathcal{F}_{\mathcal{R}})$, such that the decision a amounts to choosing a particular distribution P_a on the rewards.

Example 3.2 (Rock paper scissors) Consider the simple game of rock paper scissors, where your opponent plays a move at the same time as you, so that you cannot influence his move. The opponent's moves are $\Omega = \{\omega_{\mathrm{R}}, \omega_{\mathrm{P}}, \omega_{\mathrm{S}}\}$ for *rock*, *paper*, *scissors*, respectively, which also corresponds to your decision set $\mathcal{A} = \{a_{\mathrm{R}}, a_{\mathrm{P}}, a_{\mathrm{S}}\}$. The reward set is $\mathcal{R} = \{\text{Win}, \text{Draw}, \text{Lose}\}$.

You have studied your opponent for some time and you *believe* that he is most likely to play rock $P(\omega_{\mathrm{R}}) = 3/6$, somewhat likely to play paper $P(\omega_{\mathrm{P}}) = 2/6$, and less likely to play scissors: $P(\omega_{\mathrm{S}}) = 1/6$.

What is the probability of each reward, for each decision you make? Taking the example of a_{R}, we see that you win if the opponent plays *scissors* with probability $1/6$, you lose if the opponent plays *paper* ($2/6$), and you draw if he plays *rock* ($3/6$). Consequently, we can convert the outcome probabilities to reward probabilities for every decision:

$$\begin{aligned}
&P_{a_{\mathrm{R}}}(\text{Win}) = 1/6, && P_{a_{\mathrm{R}}}(\text{Draw}) = 3/6, && P_{a_{\mathrm{R}}}(\text{Lose}) = 2/6,\\
&P_{a_{\mathrm{P}}}(\text{Win}) = 3/6, && P_{a_{\mathrm{P}}}(\text{Draw}) = 2/6, && P_{a_{\mathrm{P}}}(\text{Lose}) = 1/6,\\
&P_{a_{\mathrm{S}}}(\text{Win}) = 2/6, && P_{a_{\mathrm{S}}}(\text{Draw}) = 1/6, && P_{a_{\mathrm{S}}}(\text{Lose}) = 3/6.
\end{aligned}$$

Of course, what you play depends on our own utility function. If you prefer winning over drawing or losing, you could for example have the utility function $U(\text{Win}) = 1$, $U(\text{Draw}) = 0$, $U(\text{Lose}) = -1$. Then, since $\mathbb{E}_a U = \sum_{\omega \in \Omega} U(\omega, a) P_a(\omega)$, we have

$$\begin{aligned}
E_{a_{\mathrm{R}}} U &= -1/6,\\
E_{a_{\mathrm{P}}} U &= 2/6,\\
E_{a_{\mathrm{S}}} U &= -1/6,
\end{aligned}$$

so that based on your belief, choosing paper is best.

The above example illustrates that every decision that we make creates a corresponding probability distribution on the rewards. While the outcome of the experiment is independent of the decision, the distribution of rewards is effectively chosen by our decision.

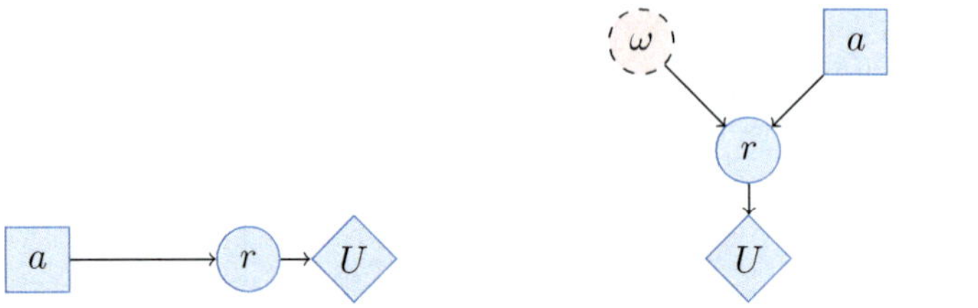

(a) The combined decision problem

(b) The separated decision problem

Fig. 3.1 Decision diagrams for the combined and separated formulation of the decision problem. Squares denote decision variables, diamonds denote utilities. All other variables are denoted by circles. Arrows denote the flow of dependency

Expected utility
The expected utility of any decision $a \in \mathcal{A}$ under P is:

$$\mathbb{E}_{P_a}(U) = \int_{\mathcal{R}} U(r)\,\mathrm{d}P_a(r) = \int_{\Omega} U[\rho(\omega, a)]\,\mathrm{d}P(\omega)$$

From now on, we shall use the simple notation

$$U(P, a) \triangleq \mathbb{E}_{P_a} U$$

to denote the expected utility of a under distribution P.

Instead of viewing the decision as effectively choosing a distribution over rewards (Fig. 3.1a) we can separate the random part of the process from the deterministic part (Fig. 3.1b) by considering a measure P on the space of outcomes Ω, such that the reward depends on both a and the outcome $\omega \in \Omega$ through the reward function $\rho(\omega, a)$. The optimal decision is of course always the $a \in \mathcal{A}$ maximizing $\mathbb{E}(U \mid P_a)$. However, this structure allows us to clearly distinguish the controllable from the random part of the rewards.

The probability measure induced by decisions
For every $a \in \mathcal{A}$, the function $\rho : \Omega \times \mathcal{A} \to \mathcal{R}$ induces a probability distribution P_a on $\mathcal{R}$. In fact, for any $B \in \mathcal{F}_{\mathcal{R}}$:

$$P_a(B) \triangleq \mathbb{P}(\rho(\omega, a) \in B) = P(\omega \,|\rho(\omega, a) \in B)$$

Table 3.1 Rewards, utilities, expected utility for 20% probability of rain

$\rho(\omega, a)$	a_1	a_2
ω_1	Dry, carrying umbrella	Wet
ω_2	Dry, carrying umbrella	Dry
$U[\rho(\omega, a)]$	a_1	a_2
ω_1	–1	–10
ω_2	–1	0
$\mathbb{E}_P(U \mid a)$	–1	–2

The above equation requires that the following technical assumption is satisfied. As usual, we employ the expected utility hypothesis (Assumption 2.3.4). Thus, we should choose the decision that results in the highest expected utility.

Assumption 3.2.3 The sets $\{\omega \mid \rho(\omega, a) \in B\}$ belong to $\mathcal{F}_\Omega$. That is, ρ is $\mathcal{F}_\Omega$-measurable for any a.

The dependency structure of this problem in either formulation can be visualized in the *decision diagram* shown in Fig. 3.1.

Example 3.3 (Continuation of Example 3.1) You are going to work, and it might rain. The forecast said that the probability of rain (ω_1) was 20%. What do you do?

- a_1: Take the umbrella.
- a_2: Risk it!

The reward of a given outcome and decision combination, as well as the respective utility is given in Table 3.1.

3.2.2 Decision Diagrams

Decision diagrams, also known as *decision networks* or *influence diagrams*, are used to show dependencies between different variables. As illustrated in the examples shown in Fig. 3.1, if an arrow points from a variable x to a variable y it means that y depends on x. In other words, y has x as an input. In general, decision diagrams include the following types of nodes:

- *Choice nodes* (denoted by squares) are nodes whose values can be directly chosen by the decision maker. In general there may be more than one decision maker involved.
- *Value nodes* (denoted by diamonds) are the nodes that the decision maker is interested in influencing. That is, the utility of the decision maker is always a function of the value nodes.
- *Circle nodes* are used to denote all other types of variables. These include deterministic or stochastic variables.

- *Line style and colour* represent quantities that part of the problem, but are not observed by one or more of the decision makers. These are usually called *latent variables*.

Let us take a look at the example of Fig. 3.1b, the reward is a function of both ω and a, i.e., $r = \rho(\omega, a)$, while ω depends only on the probability distribution P. Typically, there must be a path from a choice node to a value node, otherwise nothing the decision maker can do will influence the utility. Nodes belonging to or observed by different players will usually be denoted by different lines or colors. In Fig. 3.1b, ω, which is not observed, is shown with a dashed line.

3.2.3 Statistical Estimation*

Statistical decision problems arise particularly often in *parameter estimation,* such as estimating the covariance matrix of a Gaussian random variable. In this setting, the unknown outcome of the experiment ω is called a *parameter*, while the set of outcomes Ω is called the *parameter space*. Classical statistical estimation involves selecting a single parameter value on the basis of observations. This requires us to specify a preference for different types of estimation errors, and is distinct from the standard Bayesian approach to estimation, which calculates a full distribution over all possible parameters.

A simple example is estimating the distribution of votes in an election from a small sample. Depending on whether we are interested in predicting the vote share of individual parties or the most likely winner of the election, we can use a distribution over vote shares (possibly estimated through standard Bayesian methodology) to decide on a share or the winner.

Example 3.4 (Voting) Assume you wish to estimate the number of votes for different candidates in an election. The *unknown parameters* of the problem mainly include: the percentage of likely voters in the population, the probability that a likely voter is going to vote for each candidate. One simple way to estimate this is by polling.

Consider a nation with k political parties. Let $\omega = (\omega_1, \ldots, \omega_k) \in [0, 1]^k$ be the voting proportions for each party. We wish to make a guess $a \in [0, 1]^k$. How should we guess, given a distribution $P(\omega)$? How should we select U and ρ? This depends on what our goal is, when we make the guess.

If we wish to give a reasonable estimate about the votes of all the k parties, we can use the squared error: First, set the error vector $r = (\omega_1 - a_1, \ldots, \omega_k - a_k) \in [0, 1]^k$. Then we set $U(r) \triangleq -\|r\|^2$, where $\|r\|^2 = \sum_i |\omega_i - a_i|^2$.

If on the other hand, we just want to predict the winner of the election, then the actual percentages of all individual parties are not important. In that case, we can set $r = 1$ if $\arg\max_i \omega_i = \arg\max_i a_i$ and 0 otherwise, and $U(r) = r$.

Losses and risks
In such problems, it is common to specify a loss instead of a utility. This is usually the negative utility:

Definition 3.2.2 (*Loss*)

$$\ell(\omega, a) = -U[\rho(\omega, a)]$$

Given the above, instead of the expected utility, we consider the expected loss, or risk.

Definition 3.2.3 (*Risk*)

$$\kappa(P, a) = \int_{\Omega} \ell(\omega, a) \, \mathrm{d}P(\omega)$$

Of course, the optimal decision is a minimizing κ.

3.3 Bayes Decisions

The decision which maximizes the expected utility under a particular distribution P, is called the *Bayes-optimal* decision, or simply the *Bayes decision*. The probability distribution P is supposed to reflect all our uncertainty about the problem. Note that in the following we usually drop the reward function ρ from the decision problem and consider utility functions U that map directly from $\Omega \times \mathcal{A}$ to $\mathbb{R}$.

Definition 3.3.1 (*Bayes-optimal utility*) Consider an outcome (or parameter) space Ω, a decision space $\mathcal{A}$, and a utility function $U : \Omega \times \mathcal{A} \to \mathbb{R}$. For any probability distribution P on Ω, the Bayes-optimal utility $U^*(P)$ is defined as the smallest upper bound on $U(P, a)$ over all decisions $a \in \mathcal{A}$. That is,

$$U^*(P) = \sup_{a \in \mathcal{A}} U(P, a). \tag{3.3.1}$$

The maximization over decisions is usually not easy. However, there exist a few cases where it is relatively simple. The first of those is when the utility function is the negative squared error.

Example 3.5 (Quadratic loss) Consider $\Omega = \mathbb{R}^k$ and $\mathcal{A} = \mathbb{R}^k$. The utility function that, for any point $\omega \in \mathbb{R}$, is defined as

$$U(\omega, a) = -\|\omega - a\|^2$$

is called quadratic loss.

Quadratic loss is a very important special case of utility functions, as it is easy to calculate the optimal solution. This is illustrated by the following theorem.

Theorem 3.3.1 *Let P be a measure on Ω and $U : \Omega \times \mathcal{A} \to \mathbb{R}$ be the quadratic loss defined in Example 3.5. Then the decision*

$$a = \mathbb{E}_P(\omega)$$

maximizes the expected utility $U(P, a)$, under the technical assumption that $\partial/\partial a|\omega - a|^2$ is measurable with respect to $\mathcal{F}_{\mathbb{R}}$.

Proof The expected utility of decision a is given by

$$U(P, a) = -\int_\Omega \|\omega - a\|^2 \, \mathrm{d}P(\omega).$$

Taking derivatives, due to the measurability assumption, we can swap the order of differentiation and integration and obtain

$$\begin{aligned}
\frac{\partial}{\partial a}\int_\Omega \|\omega - a\|^2 \, \mathrm{d}P(\omega) &= \int_\Omega \frac{\partial}{\partial a}\|\omega - a\|^2 \, \mathrm{d}P(\omega) \\
&= 2\int_\Omega (a - \omega) \, \mathrm{d}P(\omega) \\
&= 2\int_\Omega a \, \mathrm{d}P(\omega) - 2\int_\Omega \omega \, \mathrm{d}P(\omega) \\
&= 2a - 2\mathbb{E}(\omega).
\end{aligned}$$

Setting the derivative equal to 0 and noting that the utility is concave, we see that the expected utility is maximized for $a = \mathbb{E}_P(\omega)$. □

Another simple example is the absolute error, where $U(\omega, a) = |\omega - a|$. The solution in this case differs significantly from the squared error. As can be seen from Fig. 3.2a, for absolute loss, the optimal decision is to choose the a that is closest to the most likely ω. Figure 3.2b illustrates the finding of Theorem 3.3.1.

*3.3.1 Convexity of the Bayes-Optimal Utility**

Although finding the optimal decision for an arbitrary utility U and distribution P may be difficult, fortunately the Bayes-optimal utility has some nice properties which enable it to be approximated rather well. In particular, for any decision, the expected utility is linear with respect to our belief P. Consequently, the Bayes-optimal utility is convex with respect to P. This firstly implies that there is a unique "worst" distribution P, against which we cannot do very well. Secondly, we can approximate

Fig. 3.2 Expected utility curves for different values of $P(\omega = 0)$ in $\{0.1, 0.25, 0.5, 0.75\}$, as the decision a varies in $[0, 1]$

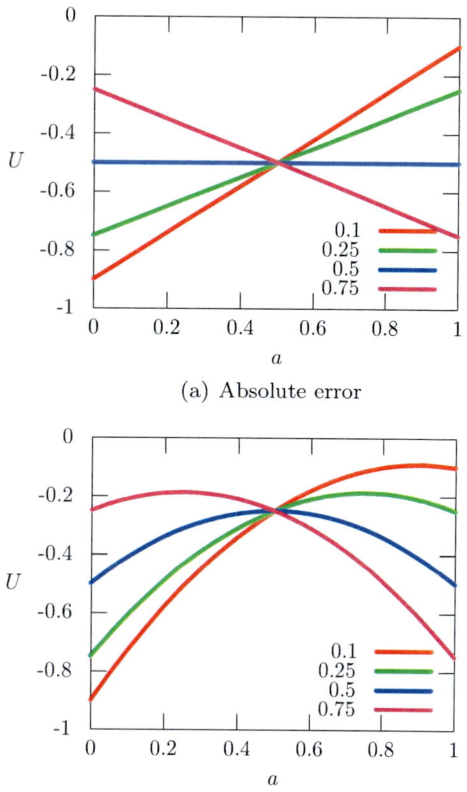

(a) Absolute error

(b) Quadratic error

the Bayes-utility very well for all possible distributions by generalizing from a small number of distributions. In order to define linearity and convexity, we first introduce the concept of a mixture of distributions.

Consider two probability measures P, Q on $(\Omega, \mathcal{F}_\Omega)$. These define two alternative distributions for ω. For any P, Q and $\alpha \in [0, 1]$, we define the *mixture of distributions*

mixture of distributions

$$Z_\alpha \triangleq \alpha P + (1 - \alpha) Q \tag{3.3.2}$$

to mean the probability measure such that $Z_\alpha(A) = \alpha P(A) + (1 - \alpha) Q(A)$ for any $A \in \mathcal{F}_\Omega$. For any fixed choice a, the expected utility varies linearly with α:

Remark 3.3.1 (*Linearity of the expected utility*) If Z_α is as defined in (3.3.2), then, for any $a \in \mathcal{A}$,

$$U(Z_\alpha, a) = \alpha\, U(P, a) + (1 - \alpha)\, U(Q, a).$$

Proof

$$
\begin{aligned}
U(Z_\alpha, a) &= \int_\Omega U(\omega, a)\, \mathrm{d}Z_\alpha(\omega) \\
&= \alpha \int_\Omega U(\omega, a)\, \mathrm{d}P(\omega) + (1-\alpha) \int_\Omega U(\omega, a)\, \mathrm{d}Q(\omega) \\
&= \alpha\, U(P, a) + (1-\alpha)\, U(Q, a).
\end{aligned}
$$
□

However, if we consider Bayes-optimal decisions, this is no longer true, because the optimal decision depends on the distribution. In fact, the utility of Bayes-optimal decisions is convex, as the following theorem shows.

Theorem 3.3.2 *For probability measures P, Q on Ω and any $\alpha \in [0, 1]$,*

$$U^*[Z_\alpha] \le \alpha\, U^*(P) + (1-\alpha)\, U^*(Q).$$

Proof From the definition of the expected utility (3.3.1), for any decision $a \in \mathcal{A}$,

$$U(Z_\alpha, a) = \alpha\, U(P, a) + (1-\alpha)\, U(Q, a).$$

Hence, by definition (3.3.1) of the Bayes-utility, we have

$$
\begin{aligned}
U^*(Z_\alpha) &= \sup_{a \in \mathcal{A}} U(Z_\alpha, a) \\
&= \sup_{a \in \mathcal{A}} [\alpha\, U(P, a) + (1-\alpha)\, U(Q, a)].
\end{aligned}
$$

As $\sup_x [f(x) + g(x)] \le \sup_x f(x) + \sup_x g(x)$, we obtain

$$
\begin{aligned}
U^*[Z_\alpha] &\le \alpha \sup_{a \in \mathcal{A}} U(P, a) + (1-\alpha) \sup_{a \in \mathcal{A}} U(Q, a) \\
&= \alpha\, U^*(P) + (1-\alpha)\, U^*(Q).
\end{aligned}
$$
□

As we have proven, the expected utility is linear with respect to Z_α. Thus, for any fixed action a we obtain a line as those shown in Fig. 3.3. By Theorem 3.3.2, the Bayes-optimal utility is convex. Furthermore, the minimizing decision for any Z_α is tangent to the Bayes-optimal utility at the point $(Z_\alpha, U^*(Z_\alpha))$. If we take a decision that is optimal with respect to some Z, but the distribution is in fact $Q \neq Z$, then we are not far from the optimal, if Q, Z are close and U^* is smooth. Consequently, we can trivially lower bound the Bayes utility by examining any arbitrary finite set of decisions $\hat{\mathcal{A}} \subseteq \mathcal{A}$. That is,

$$U^*(P) \ge \max_{a \in \hat{\mathcal{A}}} U(P, a)$$

for any probability distribution P on Ω. In addition, we can upper-bound the Bayes utility as follows. Take any two distributions P_1, P_2 over Ω. Then, the upper bound

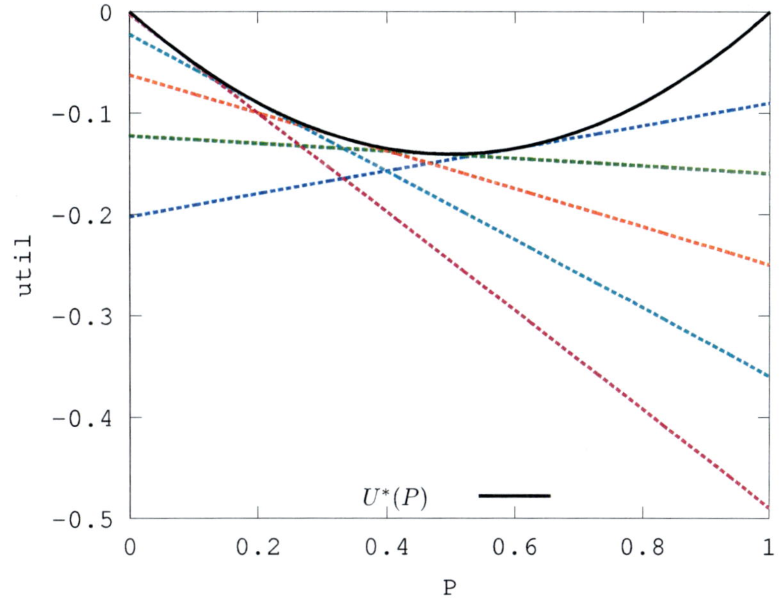

Fig. 3.3 A strictly convex Bayes utility

$$U^*(\alpha P_1 + (1-\alpha)P_2) \leq \alpha\, U^*(P_1) + (1-\alpha)\, U^*(P_2)$$

holds due to convexity. The two bounds suggest an algorithm for successive approximation of the Bayes-optimal utility, by looking for the largest gap between the lower and the upper bounds.

3.4 Statistical and Strategic Decision Making

In this section we consider more general strategies that do not simply pick a single fixed decision. Further, we will consider other criteria than maximizing expected utility and minimizing risk, respectively.

Strategies Instead of choosing a specific decision, we could instead choose to randomize our decision somehow. In other words, instead of our choices being specific decisions, we can choose among distributions over decisions. For example, instead of choosing to eat lasanga or beef, we choose to throw a coin and eat lasagna if the coin comes heads and beef otherwise. Accordingly, in the following we will consider strategies that are probability measures on $\mathcal{A}$, the set of which we will denote by $\mathbb{\Delta}(\mathcal{A})$.

Definition 3.4.1 (*Strategy*) A strategy $\sigma \in \mathbb{\Delta}(\mathcal{A})$ is a probability distribution over $\mathcal{A}$ and determines for each a in $\mathcal{A}$ the probability with which a is chosen.

Interestingly, *for the type of problems that we have considered so far*, even if we expand our choices to the set of all possible probability measures on $\mathcal{A}$, there always is one decision (rather than a strategy) which is optimal.

Theorem 3.4.1 *Consider any statistical decision problem with probability measure P on outcomes Ω and with utility function $U : \Omega \times \mathcal{A} \to \mathbb{R}$. Further let $a^* \in \mathcal{A}$ such that $U(P, a^*) \geq U(P, a)$ for all $a \in \mathcal{A}$. Then for any probability measure σ on $\mathcal{A}$,*

$$U(P, a^*) \geq U(P, \sigma).$$

Proof

$$\begin{aligned} U(P, \sigma) &= \int_{\mathcal{A}} U(P, a) \, \mathrm{d}\sigma(a) \\ &\leq \int_{\mathcal{A}} U(P, a^*) \, \mathrm{d}\sigma(a) \\ &= U(P, a^*) \int_{\mathcal{A}} \mathrm{d}\sigma(a) \\ &= U(P, a^*) \end{aligned}$$

□

This theorem should be not be applied naively. It only states that if we know P then the expected utility of the best fixed/deterministic decision $a^* \in \mathcal{A}$ cannot be increased by randomizing between decisions. For example, it does not make sense to apply this theorem to cases where P itself is completely or partially unknown (e.g., when P is chosen by somebody else and its value remains hidden to us).

3.4.1 Alternative Notions of Optimality

There are some situations where maximizing expected utility with respect to the distribution on outcomes is unnatural. Two simple examples are the following.

Maximin and minimax policies If there is no information about ω available, it may be reasonable to take a pessimistic approach and select a^* that maximizes the utility in the worst-case ω. The respective *maximin* value

maximin

$$U_* = \max_a \min_\omega U(\omega, a) = \min_\omega U(\omega, a^*)$$

can essentially be seen as how much utility we would be able to obtain, if we were to make a decision a first, and nature were to select an adversarial decision ω later.

On the other hand, the *minimax* value is minimax

$$U^* = \min_\omega \max_a U(\omega, a) = \max_a U(\omega^*, a),$$

where $\omega^* \triangleq \arg\min_\omega \max_a U(\omega, a)$ is the worst-case choice nature could make, if we were to select our own decision a after its own choice was revealed to us.

To illustrate this, let us consider the following example.

Example 3.6 You consider attending an open air concert. The weather forecast reports 50% probability of rain. Going to the concert (a_1) will give you a lot of pleasure if it doesn't rain (ω_2), but in case of rain (ω_1) you actually would have preferred to stay at home (a_2). Since in general you prefer nice weather to rain also in case you decided not to go you prefer ω_2 to ω_1. The reward of a given outcome-decision combination, as well as the respective utility is given in Table 3.2. We see that a_1 maximizes expected utility. However, under a worst-case assumption this is not the case, i.e., the maximin solution is a_2.

Note that by definition

$$U^* \geq U(\omega^*, a^*) \geq U_*. \tag{3.4.1}$$

Maximin/minimax problems are a special case of problems in game theory, in particular two-player zero-sum games. The minimax problem can be seen as a game where the maximizing player plays first, and the minimizing player second. If $U^* = U_*$, then the game is said to *have a value*, which implies that if both players are playing optimal, then it doesn't matter which player moves first. More details about these types of problems will be given in Sect. 3.4.2.

Regret Instead of calculating the expected utility for each possible decision, we could instead calculate how much utility we would have obtained if we had made the best decision in hindsight. Consider, for example the problem in Table 3.2. There the optimal action is either a_1 or a_2, depending on whether we accept the probability P over Ω, or adopt a worst-case approach. However, after we make a specific decision, we can always look at the best decision we could have made given the actual outcome ω.

Definition 3.4.2 (*Regret*) The regret of σ is how much we lose compared to the best decision in hindsight, that is,

$$L(\omega, a) \triangleq \max_{a'} U(\omega, a') - U(\omega, a).$$

As an example let us revisit Example 3.6. Given the regret of each decision-outcome pair, we can determine the decision minimizing expected regret $\mathbb{E}(L \mid P, a)$ and minimizing maximum regret $\max_\omega L(\omega, a)$, analogously to expected utility and minimax utility. Table 3.3 shows that the choice minimizing regret either in expectation or in the minimax sense is a_1 (going to the concert). Note that this is a different outcome than before when we considered utility, which shows that the concept of regret may result in different decisions.

Table 3.2 Utility function, expected utility and maximin utility of Example 3.6

$U(\omega, a)$	a_1	a_2
ω_1	–1	0
ω_2	10	1
$\mathbb{E}(U \mid P, a)$	*4.5*	–0.5
$\min_\omega U(\omega, a)$	–1	*0*

Table 3.3 Regret, in expectation and minimax for Example 3.6

$L(\omega, a)$	a_1	a_2
ω_1	1	0
ω_2	0	9
$\mathbb{E}(L \mid P, a)$	*0.5*	4.5
$\max_\omega L(\omega, a)$	*1*	9

*3.4.2 Solving Minimax Problems**

We now view minimax problems as two player games, where one player chooses a and the other player chooses ω. The decision diagram for this problem is given in Fig. 3.4, where the dashed line indicates that, from the point of view of the decision maker, nature's choice is unobserved before she makes her own decision. A simultaneous two-player game is a game where both players act without knowing each other's decision, see Fig. 3.5. From the point of view of the player that chooses a, this is equivalent to assuming that ω is hidden, as shown in Fig. 3.4. There are other variations of such games, however. For example, their moves may be revealed after they have played. This is important in the case where the game is played *repeatedly*. However, what is usually revealed is not the belief ξ, which is something assumed to be internal to player one, but ω, the actual decision made by the first player. In other cases, we might have that U itself is not known, and we only observe $U(\omega, a)$ for the choices made.

Minimax utility, regret and loss In the following we again consider strategies as defined in Definition 3.4.1. If the decision maker knows the outcome, then the additional flexibility by randomizing over the actions does not help. As we showed

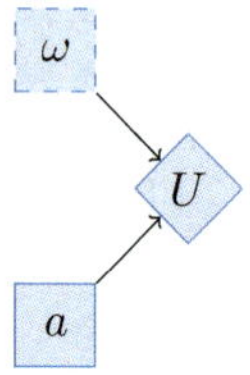

Fig. 3.4 Simultaneous two-player stochastic game. The first player (nature) chooses ω, and the second player (the decision maker) chooses a. Then the second player obtains utility $U(\omega, a)$

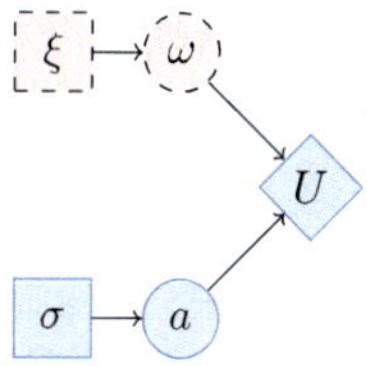

Fig. 3.5 Simultaneous two-player stochastic game. The first player (nature) chooses ξ, and the second player (the decision maker) chooses σ. Then $\omega \sim \xi$ and $a \sim \sigma$ and the second player obtains utility $U(\omega, a)$

for the general case of a distribution over Ω, a simple decision is as good as any randomized strategy:

Remark 3.4.1 For each ω, there is some a such that

$$U(\omega, a) \in \max_{\sigma \in \triangle \mathcal{A}} U(\omega, \sigma). \tag{3.4.2}$$

What follows are some rather trivial remarks connecting regret with utility in various cases.

Remark 3.4.2

$$L(\omega, \sigma) = \sum_a \sigma(a) L(w, a) \geq 0,$$

with equality iff σ is ω-optimal.

Proof

$$L(\omega, \sigma) = \max_{\sigma'} U(\omega, \sigma') - U(\omega, \sigma) = \max_{\sigma'} U(\omega, \sigma') - \sum_a \sigma(a) U(\omega, a)$$
$$= \sum_a \sigma(a) \big(\max_{\sigma'} U(\omega, \sigma') - U(\omega, a)\big) \geq 0$$

The equality in case of optimality is obvious. □

Remark 3.4.3

$$L(\omega, \sigma) = \max_a U(\omega, a) - U(\omega, \sigma)$$

Proof As (3.4.2) shows, for any fixed ω, the best decision is always deterministic, so that

$$\sum_{a'} \sigma(a') L(\omega, a') = \sum_{a'} \sigma(a') [\max_{a \in \mathcal{A}} U(\omega, a) - U(\omega, a')]$$
$$= \max_{a \in \mathcal{A}} U(\omega, a) - \sum_{a'} \sigma(a') U(\omega, a').$$ □

Table 3.4 Even-bet utility

U	ω_1	ω_2
a_1	1	−1
a_2	0	0

Table 3.5 Even-bet regret

L	ω_1	ω_2
a_1	0	1
a_2	1	0

Remark 3.4.4 $L(\omega, \sigma) = -U(\omega, \sigma)$ iff $\max_a U(\omega, a) = 0$.

Proof If

$$\max_{\sigma'} U(\omega, \sigma') - U(\omega, \sigma) = -U(\omega, \sigma)$$

then

$$\max_{\sigma'} U(\omega, \sigma') = \max_a U(\omega, a) = 0.$$

The converse holds as well, which finishes the proof. □

The following example demonstrates that in general the minimax regret will be achieved by a randomized strategy.

Example 3.7 (An even-money bet) Consider the decision problem described in Table 3.4. The respective loss is given in Table 3.5. The maximum regret of a strategy σ can be written as

$$\begin{aligned}\max_\omega L(\omega, \sigma) &= \max_\omega \sum_a \sigma(a) L(\omega, a) \\ &= \max_\omega \sigma(a) \mathbb{I}\{a = a_i \wedge \omega \neq \omega_i\} \cdot 1,\end{aligned}$$

since $L(a, \omega) = 0$ when $a = a_i$ and $\omega \neq \omega_i$. Note that $\max_\omega L(\omega, \sigma) \geq \frac{1}{2}$ and that equality is obtained iff $\sigma(a) = \frac{1}{2}$, giving minimax regret $L^* = \frac{1}{2}$.

3.4.3 Two-Player Games

In this section we give a few more details about the connections between minimax theory and the theory of two-player games. In particular, we extend the actions of nature to $\Delta(\Omega)$, the probability distributions over Ω and as before consider strategies in $\Delta(\mathcal{A})$.

For two distributions σ, ξ on $\mathcal{A}$ and Ω, we define our expected utility as

$$U(\xi, \sigma) \triangleq \sum_{\omega \in \Omega} \sum_{a \in \mathcal{A}} U(\omega, a)\xi(\omega)\sigma(a).$$

Then we define the maximin policy σ^* to satisfy

$$\min_{\xi} U(\xi, \sigma^*) = U_* \triangleq \max_{\sigma} \min_{\xi} U(\xi, \sigma).$$

The minimax prior ξ^* satisfies

$$\max_{\sigma} U(\xi^*, \sigma) = U^* \triangleq \min_{\xi} \max_{\sigma} U(\xi, \sigma),$$

where the solution exists as long as $\mathcal{A}$ and Ω are finite, which we will assume in the following.

Expected regret
We can now define the expected regret for a given pair of distributions ξ, σ as

$$\begin{aligned} L(\xi, \sigma) &= \max_{\sigma'} \sum_{\omega} \xi(\omega) \left\{U(\omega, \sigma') - U(\omega, \sigma)\right\} \\ &= \max_{\sigma'} U(\xi, \sigma') - U(\xi, \sigma). \end{aligned}$$

Not all minimax and maximin policies result in the same value. The following theorem gives a condition under which the game does have a value.

Theorem 3.4.2 *If there exist distributions*[1] ξ^*, σ^* *and* $C \in \mathbb{R}$ *such that*

$$U(\xi^*, \sigma) \leq C \leq U(\xi, \sigma^*) \qquad \forall \xi, \sigma$$

then

$$U^* = U_* = U(\xi^*, \sigma^*) = C.$$

Proof Since $C \leq U(\xi, \sigma^*)$ for all ξ we have

$$C \leq \min_{\xi} U(\xi, \sigma^*) \leq \max_{\sigma} \min_{\xi} U(\xi, \sigma) = U_*.$$

Similarly,

$$C \geq \max_{\sigma} U(\xi^*, \sigma) \geq \min_{\xi} \max_{\sigma} U(\xi, \sigma) = U^*.$$

[1] These distributions may be *singular*, that is, they may be concentrated in one point. For example, σ^* is singular, if $\sigma^*(a) = 1$ for some a and $\sigma^*(a') = 0$ for all $a' \neq a$.

By (3.4.1) it follows that

$$C \geq U^* \geq U_* \geq C. \qquad \square$$

Theorem 3.4.2 gives a sufficient condition for a game having a value. In fact, the type of games we have been looking at so far are called bilinear games. For these, a solution always exists and there are efficient methods for finding it.

Definition 3.4.3 A bilinear game is a tuple $(U, \Xi, \Sigma, \Omega, \mathcal{A})$ with $U : \Xi \times \Sigma \to \mathbb{R}$ such that all $\xi \in \Xi$ are arbitrary distributions on Ω and all $\sigma \in \Sigma$ are arbitrary distributions on $\mathcal{A}$ with

$$U(\xi, \sigma) \triangleq \mathbb{E}(U \mid \xi, \sigma) = \sum_{\omega, a} U(\omega, a)\, \sigma(a)\, \xi(\omega).$$

Theorem 3.4.3 *For a bilinear game, $U^* = U_*$. In addition, the following three conditions are equivalent:*

1. *σ^* is maximin, ξ^* is minimax, and $U^* = C$.*
2. *$U(\xi, \sigma^*) \geq C \geq U(\xi^*, \sigma)$ for all ξ, σ.*
3. *$U(\omega, \sigma^*) \geq C \geq U(\xi^*, a)$ for all ω, a.*

3.4.3.1 Linear Programming Formulation

While general games may be hard, bilinear games are easy, in the sense that minimax solutions can be found with well-known algorithms. One such method is linear programming. The problem

$$\max_{\sigma} \min_{\xi} U(\xi, \sigma),$$

where ξ, σ are distributions over finite domains, can be converted to finding σ corresponding to the greatest lower bound $v_\sigma \in \mathbb{R}$ on the utility. Using matrix notation, set $\boldsymbol{U}$ to be the matrix such that $\boldsymbol{U}_{\omega,a} = U(\omega, a)$, and consider the vectors $\sigma_a = \sigma(a)$ and $\boldsymbol{\xi}_\omega = \xi(\omega)$. Then the problem can be written as:

$$\max \left\{ v_\sigma \mid (\boldsymbol{U}\sigma)_j \geq v_\sigma \forall j,\ \sum_i \sigma_i = 1,\ \sigma_i \geq 0 \,\forall i \right\}$$

Equivalently, we can find ξ with the least upper bound:

$$\min \left\{ v_\xi \mid (\boldsymbol{\xi}^\top \boldsymbol{U})_i \leq v_\xi \forall i,\ \sum_j \xi_j = 1,\ \xi_j \geq 0 \,\forall j \right\},$$

where everything has been written in matrix form. In fact, one can show that $v_\xi = v_\sigma$, thus obtaining Theorem 3.4.3.

To understand the connection of two-person games with Bayesian decision theory, take another look at Fig. 3.3, seeing the risk as negative expected utility, or as the opponent's gain. Each of the decision lines represents nature's gain as she chooses different prior distributions, while we keep our policy σ fixed. The bottom horizontal line that would be tangent to the Bayes-optimal utility curve would be minimax: if nature were to change priors, it would not increase its gain, since the line is horizontal. On the other hand, if we were to choose a different tangent line, we would only increase nature's gain (and decrease our utility).

3.5 Decision Problems with Observations

So far we have only examined problems where the outcomes were drawn from some fixed distribution. This distribution constituted our subjective belief about what the unknown parameter is. Now, we examine the case where we can obtain some observations that depend on the unknown ω before we make our decision, cf. Fig. 3.6. These observations should give us more information about ω before making a decision. Intuitively, we should be able to make decisions by simply considering the posterior distribution.

In this setting, we once more aim to take some decision $a \in \mathcal{A}$ so as to maximize expected utility. As before, we have a prior distribution ξ on some parameter $\omega \in \Omega$, representing what we know about ω. Consequently, the expected utility of any fixed decision a is going to be $\mathbb{E}_\xi(U \mid a)$.

However, now we may obtain more information about ω before making a final decision. In particular, each ω corresponds to a *model* of the world P_ω, which is a probability distribution over some observation space $\mathcal{S}$, such that $P_\omega(X)$ is the probability that the observation is in $X \subset \mathcal{S}$. The set of parameters Ω thus defines a family of models

$$\mathscr{P} \triangleq \{P_\omega \mid \omega \in \Omega\}.$$

Now, consider the case where we take an observation x from the true model P_{ω^*} before making a decision. We can represent the dependency of our decision on the observation by making our decision a function of x.

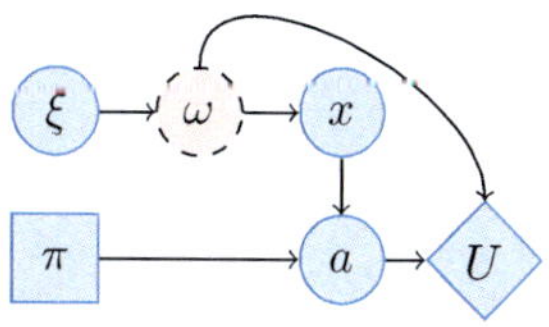

Fig. 3.6 Statistical decision problem with observations

Definition 3.5.1 (*Policy*) A policy $\pi : \mathcal{S} \to \mathcal{A}$ maps any observation to a decision.[2]

Given a policy π, its expected utility is given by

$$U(\xi, \pi) \triangleq \mathbb{E}_\xi \{U[\omega, \pi(x)]\} = \int_\Omega \left(\int_\mathcal{S} U[\omega, \pi(x)] \, \mathrm{d}P_\omega(x) \right) \mathrm{d}\xi(\omega).$$

This is the standard Bayesian framework for decision making. It may be slightly more intuitive in some case to use the notation $\psi(x \mid \omega)$, in order to emphasize that this is a conditional distribution. However, there is no technical difference between the two notations.

When the set of policies includes all constant policies, then there is a policy π^* at least as good as the best fixed decision a^*. This is formalized in the following remark.

Remark 3.5.1 Let Π denote a set of policies $\pi : \mathcal{S} \to \mathcal{A}$. If for each $a \in \mathcal{A}$ there is a $\pi \in \Pi$ such that $\pi(x) = a \; \forall x \in \mathcal{S}$, then $\max_{\pi \in \Pi} \mathbb{E}_\xi(U \mid \pi) \geq \max_{a \in \mathcal{A}} \mathbb{E}_\xi(U \mid a)$.

Proof The proof follows by setting Π_0 to be the set of constant policies. The result follows since $\Pi_0 \subset \Pi$. □

We conclude this section with a simple example about deciding whether or not to go to a restaurant, given some expert opinions.

Example 3.8 Consider the problem of deciding whether or not to go to a particular restaurant. Let $\Omega = [0, 1]$ with $\omega = 0$ meaning the food is in general horrible and $\omega = 1$ meaning the restaurant is great. Let $x_1, \ldots, x_n$ be n expert opinions in $\mathcal{S} = \{0, 1\}$ about the restaurant, where 1 means that the restaurant is recommended by the expert and 0 means that it is not recommended. Under our model, the probability of observing $x_i = 1$ when the quality of the restaurant is ω is given by $P_\omega(1) = \omega$ and conversely $P_\omega(0) = 1 - \omega$. The probability of observing a particular sequence x of length n is[3]

$$P_\omega(x) = \omega^s (1 - \omega)^{n-s}$$

with $s = \sum_{i=1}^n x_i$.

[2] For that reason, policies are also sometimes called *decision functions* or *decision rules* in the literature.

[3] We obtain a different probability of observations under the binomial model, but the resulting posterior, and hence the policy, is the same.

3.5.1 Maximizing Utility When Making Observations

Statistical procedures based on the assumption that a distribution can be assigned to any parameter in a statistical decision problem are called *Bayesian statistical methods*. The scope of these methods has been the subject of much discussion in the statistical literature, see e.g. [1].

In the following, we shall look at different expressions for the expected utility. We shall overload the utility operator U for various cases: when the parameter is fixed, when the parameter is random, when the decision is fixed, and when the decision depends on the observation x and thus is random as well.

Expected utility of a fixed decision a with $\omega \sim \xi$
We first consider the expected utility of taking a fixed decision $a \in \mathcal{A}$, when $\mathbb{P}(\omega \in B) = \xi(B)$. This is the case we have dealt with so far with

$$U(\xi, a) \triangleq \mathbb{E}_\xi(U \mid a) = \int_\Omega U(\omega, a)\,\mathrm{d}\xi(\omega).$$

Expected utility of a policy π with fixed $\omega \in \Omega$
Next we assume that ω is fixed, but instead of selecting a decision directly, we select a decision that depends on the random observation x, which is distributed according to P_ω on $\mathcal{S}$. We do this by defining a policy $\pi : \mathcal{S} \to \mathcal{A}$. Then

$$U(\omega, \pi) = \int_{\mathcal{S}} U(\omega, \pi(x))\,\mathrm{d}P_\omega(x). \tag{3.5.1}$$

Expected utility of a policy π with $\omega \sim \xi$
Finally, we generalize to the case where ω is distributed with measure ξ. Note that the expectation of the previous expression (3.5.1) by definition can be written as

$$U(\xi, \pi) = \int_\Omega U(\omega, \pi)\,\mathrm{d}\xi(\omega), \qquad U^*(\xi) \triangleq \sup_\pi U(\xi, \pi) = U(\xi, \pi^*).$$

3.5.2 Bayes Decision Rules

We wish to construct the Bayes decision rule, that is, the policy with maximal ξ-expected utility. However, doing so by examining all possible policies is cumbersome, because (usually) there are many more policies than decisions. It is however, easy to find the Bayes decision for each possible observation. This is because it is usually possible to rewrite the expected utility of a policy in terms of the posterior distribution. While this is trivial to do when the outcome and observation spaces are finite, it can be extended to the general case as shown in the following theorem.

Theorem 3.5.1 *If U is non-negative or bounded, then we can reverse the integration order in the* normal *form*

$$U(\xi, \pi) = \mathbb{E}\{U[\omega, \pi(x)]\} = \int_{\Omega} \int_{\mathcal{S}} U[\omega, \pi(x)] \, \mathrm{d}P_{\omega}(x) \, \mathrm{d}\xi(\omega)$$

to obtain the utility in extensive *form as*

$$U(\xi, \pi) = \int_{\mathcal{S}} \int_{\Omega} U[\omega, \pi(x)] \, \mathrm{d}\xi(\omega \mid x) \, \mathrm{d}P_{\xi}(x), \tag{3.5.2}$$

where $P_{\xi}(x) = \int_{\Omega} P_{\omega}(x) \, \mathrm{d}\xi(\omega)$.

Proof To prove this when U is non-negative, we shall use Tonelli's theorem. First we need to construct an appropriate product measure. Let $p(x \mid \omega) \triangleq \frac{\mathrm{d}P_{\omega}(x)}{\mathrm{d}\nu(x)}$ be the Radon-Nikodym derivative of P_{ω} with respect to some dominating measure ν on $\mathcal{S}$. Similarly, let $p(\omega) \triangleq \frac{\mathrm{d}\xi(\omega)}{\mathrm{d}\mu(x)}$ be the corresponding derivative for ξ. Now, the utility can be written as

$$\begin{aligned} U(\xi, \pi) &= \int_{\Omega} \int_{\mathcal{S}} U[\omega, \pi(x)] \, p(x \mid \omega) \, p(\omega) \, \mathrm{d}\nu(x) \, \mathrm{d}\mu(\omega) \\ &= \int_{\Omega} \int_{\mathcal{S}} h(\omega, x) \, \mathrm{d}\nu(x) \, \mathrm{d}\mu(\omega) \end{aligned}$$

with $h(\omega, x) \triangleq U[\omega, \pi(x)] p(x \mid \omega) p(\omega)$. Clearly, if U is non-negative, then so is $h(\omega, x)$. Now we are ready to apply Tonelli's theorem to get

$$\begin{aligned}
U(\xi, \pi) &= \int_{\mathcal{S}} \int_{\Omega} h(\omega, x) \, \mathrm{d}\mu(\omega) \, \mathrm{d}\mu(x) \\
&= \int_{\mathcal{S}} \int_{\Omega} U[\omega, \pi(x)] \, p(x \mid \omega) \, p(\omega) \, \mathrm{d}\mu(\omega) \, \mathrm{d}\nu(x) \\
&= \int_{\mathcal{S}} \int_{\Omega} U[\omega, \pi(x)] \, p(\omega \mid x) \, \mathrm{d}\mu(\omega) \, p(x) \, \mathrm{d}\nu(x) \\
&= \int_{\mathcal{S}} \left[\int_{\Omega} U[\omega, \pi(x)] \, p(\omega \mid x) \, \mathrm{d}\mu(\omega) \right] \frac{\mathrm{d}P_{\xi}(x)}{\mathrm{d}\nu(x)} \, \mathrm{d}\nu(x) \\
&= \int_{\mathcal{S}} \int_{\Omega} U[\omega, \pi(x)] \, \mathrm{d}\xi(\omega \mid x) \, \mathrm{d}P_{\xi}(x).
\end{aligned}$$

When U is bounded in $[a, b]$, it suffices to consider $U' = U - a$, which is non-negative. □

We can construct an optimal policy π^* as follows. For any specific observed $x \in \mathcal{S}$, we set $\pi^*(x)$ to

$$\pi^*(x) \triangleq \arg\max_{a \in \mathcal{A}} \mathbb{E}_{\xi}(U \mid x, a) = \arg\max_{a \in \mathcal{A}} \int_{\Omega} U(\omega, a) \, \mathrm{d}\xi(\omega \mid x).$$

So now we can plug π^* in the extensive form to obtain

$$\int_{\mathcal{S}} \int_{\Omega} U[\omega, \pi^*(x)] \, \mathrm{d}\xi(\omega \mid x) \, \mathrm{d}P_{\xi}(x) = \int_{\mathcal{S}} \left(\max_{a} \int_{\Omega} U[\omega, a] \, \mathrm{d}\xi(\omega \mid x) \right) \mathrm{d}P_{\xi}(x).$$

Consequently, there is no need to completely specify the policy before we have seen x. Obviously, this would create problems when $\mathcal{S}$ is large.

Bayes' decision rule
The *optimal decision* given x is the optimal decision with respect to the *posterior* $\xi(\omega \mid x)$.

The following definitions summarize terminology that we have mostly implicitly introduced before.

Definition 3.5.2 (*Prior distribution*) The distribution ξ is called the *prior distribution* of ω.

Definition 3.5.3 (*Marginal distribution*) The distribution P_{ξ} is called the (prior) *marginal distribution* of x.

Definition 3.5.4 (*Posterior distribution*) The conditional distribution $\xi(\cdot \mid x)$ is called the *posterior distribution* of ω.

3.5.3 *Decision Problems in Classification*

Classification is the problem of deciding which class $y \in \mathcal{Y}$ some particular observation $x_t \in \mathcal{X}$ belongs to. From a decision-theoretic viewpoint, the problem can be seen at three different levels. In the first, we are given a classification model in terms of a probability distribution, and we simply we wish to classify optimally given the model. In the second, we are given a family of models, a prior distribution on the family as well as a training data set, and we wish to classify optimally according to our belief. In the last form of the problem, we are given a set of *policies* $\pi : \mathcal{X} \to \mathcal{Y}$ and we must choose the one with highest expected performance. The two latter forms of the problem are equivalent when the set of policies contains all Bayes decision rules for a specific model family.

3.5.3.1 Deciding the Class Given a Probabilistic Model

In the simple form of the problem, we are already given a classifier P that can calculate probabilities $P(y_t \mid x_t)$, and we simply must decide upon some class $a_t \in \mathcal{Y}$, so as to maximize a specific utility function. One standard utility function is the prediction accuracy

$$U_t \triangleq \mathbb{I}\{y_t = a_t\}.$$

The probability $P(y_t \mid x_t)$ is the posterior probability of the class given the observation x_t. If we wish to maximize expected utility, we can simply choose

$$a_t \in \arg\max_{a \in \mathcal{Y}} P(y_t = a \mid x_t).$$

This defines a particular, simple policy. In fact, for two-class problems with $\mathcal{Y} = \{0, 1\}$, such a rule can be often visualized as a *decision boundary* in $\mathcal{X}$, on whose one side we decide for class 0 and on whose other side for class 1.

decision boundary

3.5.3.2 Deciding the Class Given a Model Family

In the general form of the problem, we are given a *training* data set $S = \{(x_1, y_1), \ldots, (x_n, y_n)\}$, a set of *classification models* $\{P_\omega \mid \omega \in \Omega\}$, and a prior distribution ξ on Ω. For each model, we can easily calculate $P_\omega(y_1, \ldots, y_n \mid x_1, \ldots, x_n)$. Consequently, we can calculate the posterior distribution

$$\xi(\omega \mid S) = \frac{P_\omega(y_1, \ldots, y_n \mid x_1, \ldots, x_n)\, \xi(\omega)}{\sum_{\omega' \in \Omega} P_{\omega'}(y_1, \ldots, y_n \mid x_1, \ldots, x_n)\, \xi(\omega')}$$

and the posterior marginal label probability

$$P_{\xi|S}(y_t \mid x_t) \triangleq P_\xi(y_t \mid x_t, S) = \sum_{\omega \in \Omega} P_\omega(y_t \mid x_t)\, \xi(\omega \mid S).$$

We can then define the simple policy

$$a_t \in \arg\max_{a \in \mathcal{Y}} \sum_{\omega \in \Omega} P_\omega(y_t \mid x_t)\, \xi(\omega \mid S),$$

which is known as *Bayes' rule*. *Bayes' rule*

3.5.3.3 The Bayes-Optimal Policy Under Parametrization Constraints*

In some cases, we are restricted to functionally simple policies, which do not contain any Bayes rules as defined above. For example, we might be limited to linear functions of x. Let $\pi : \mathcal{X} \to \mathcal{Y}$ be such a rule and let Π be the set of allowed policies. Given a family of models and a set of training data, we wish to calculate the policy that maximizes our expected utility. For a given ω, we can indeed compute

$$U(\omega, \pi) = \sum_{x,y} U(y, \pi(x)) P_\omega(y \mid x) P_\omega(x),$$

where we assume an i.i.d. model, i.e., $x_t \mid \omega \sim P_\omega(x)$ independently of previous observations. Note that to select the optimal rule $\pi \in \Pi$ we also need to know $P_\omega(x)$. For the case where ω is unknown and we have a posterior $\xi(\omega \mid S)$, given a training data set S as before, the Bayesian framework is easily extensible and gives

$$U(\xi(\cdot \mid S), \pi) = \sum_{\omega} \xi(\omega \mid S) \sum_{x,y} U(y, \pi(x)) P_\omega(y \mid x) P_\omega(x).$$

The respective maximization is in general not trivial. However, if our policies in Π are parametrized, we can employ optimization algorithms such as gradient ascent to find a maximum. In particular, if we sample $\omega \sim \xi(\cdot \mid S)$, then

$$\nabla_\pi U(\xi(\cdot \mid S), \pi) = \sum_{x,y} \nabla_\pi U(y, \pi(x)) P_\omega(y \mid x) P_\omega(x).$$

3.5.3.4 Fairness in Classification Problems*

Any policy, when applied to large-scale, real world problems, has certain externalities. This implies that considering only the decision maker's utility is not sufficient. One such issue is fairness.

This concerns desirable properties of policies applied to a population of individuals. For example, college admissions should be decided on variables that inform us about individual merit, but fairness may also require taking into account the fact that certain communities are inherently disadvantaged. At the same time, a person should not feel that someone else in a similar situation obtained an unfair advantage. All this must be taken into account while still caring about optimizing the decision maker's utility function. As another example, consider mortgage decisions: while lenders should take into account the creditworthiness of individuals in order to make a profit, society must ensure that they do not unduly discriminate against socially vulnerable groups.

Recent work in fairness for statistical decision making in the classification setting has considered two main notions of fairness. The first uses (conditional) *independence* constraints between a sensitive variable (such as ethnicity) and other variables (such as decisions made). The second type ensures that decisions are *meritocratic*, so that better individuals are favoured. Here smoothness[4] can be used to assure that similar people are treated similarly, which helps to avoid cronyism. While a thorough discussion of fairness is beyond the scope of this book, it is useful to note that some of these concepts are impossible to strictly achieve simultaneously, but may be approximately satisfied by careful design of the policy. The recent work by [2–7], and [8] goes much more deeply on this topic.

3.5.4 Calculating Posteriors

Posterior distributions for multiple observations

We now consider how we can re-write the posterior distribution over Ω incrementally. Assume that we have a prior ξ on Ω and observe $x^n \triangleq x_1, \dots, x_n$. For the observation probability, we write:

Observation probability given history x^{n-1} and parameter ω

$$P_\omega(x_n \mid x^{n-1}) = \frac{P_\omega(x^n)}{P_\omega(x^{n-1})}$$

Then we obtain the following recursion for the posterior.

Posterior recursion

$$\xi(\omega \mid x^n) = \frac{P_\omega(x^n)\,\xi(\omega)}{P_\xi(x^n)} = \frac{P_\omega(x_n \mid x^{n-1})\xi(\omega \mid x^{n-1})}{P_\xi(x_n \mid x^{n-1})}$$

Here $P_\xi(\cdot \mid \cdot) = \int_\Omega P_\omega(\cdot \mid \cdot)\,\mathrm{d}\xi(\omega)$ is a marginal distribution.

[4] For example, through Lipschitz conditions on the policy.

Posterior distributions for multiple independent observations
Now we consider the case where, given the parameter ω, the next observation does not depend on the history: If $P_\omega(x_n \mid x^{n-1}) = P_\omega(x_n)$ then $P_\omega(x^n) = \prod_{k=1}^n P_\omega(x_k)$. Then the recursion looks as follows.

Posterior recursion with conditional independence

$$\xi_n(\omega) \triangleq \xi_0(\omega \mid x^n) = \frac{P_\omega(x^n)\,\xi_0(\omega)}{P_{\xi_0}(x_n)}$$
$$= \xi_{n-1}(\omega \mid x_n) = \frac{P_\omega(x_n)\,\xi_{n-1}(\omega)}{P_{\xi_{n-1}}(x_n)},$$

where ξ_t is the belief at time t. Here $P_{\xi_n}(\cdot \mid \cdot) = \int_\Omega P_\omega(\cdot \mid \cdot)\,\mathrm{d}\xi_n(\omega)$ is the marginal distribution with respect to the n-th posterior.

Conditional independence allows us to write the posterior update as an identical recursion at each time t. We shall take advantage of that when we look at *conjugate prior* distributions in Chap. 4. For such models, the recursion involves a particularly simple parameter update.

3.6 Summary

In this chapter, we introduced a general framework for making decisions $a \in \mathcal{A}$ whose optimality depends on an unknown outcome or parameter ω. We saw that, when our knowledge about $\omega \in \Omega$ is in terms of a probability distribution ξ on Ω, then the utility of the Bayes-optimal decision is convex with respect to ξ.

In some cases, observations $x \in \mathcal{X}$ may affect affect our belief, leading to a posterior $\xi(\cdot \mid x)$. This requires us to introduce the notion of a policy $\pi : \mathcal{X} \to \mathcal{A}$ mapping observations to decisions. While it is possible to construct a complete policy by computing $U(\xi, \pi)$ for all *policies* (normal form) and maximizing, it is frequently simpler to just wait until we observe x and compute $U[\xi(\cdot \mid x), a]$ for all *decisions* (extensive form).

In minimax settings, we can consider a fixed but unknown parameter ω or a fixed but unknown prior ξ. This links statistical decision theory to game theory.

3.7 Exercises

The first part of the exercises considers problems where we are simply given some distribution over Ω. In the second part, the distribution is a posterior distribution that depends on observations x.

3.7.1 *Problems with No Observations*

For the following exercises, we consider a set of worlds Ω and a decision set $\mathcal{A}$, as well as the following utility function $U : \Omega \times \mathcal{A} \to \mathbb{R}$:

$$U(\omega, a) = \operatorname{sinc}(\omega - a),$$

where $\operatorname{sinc}(x) = \sin(x)/x$. If ω is known and $\mathcal{A} = \Omega = \mathbb{R}$ then obviously the optimal decision is $a = \omega$, as $\operatorname{sinc}(x) \leq \operatorname{sinc}(0) = 1$. However, we consider the case where

$$\Omega = \mathcal{A} = \{-2.5, \ldots, -0.5, 0, 0.5, \ldots, 2.5\}.$$

Exercise 3.7.1 Assume ω is drawn from ξ with $\xi(\omega) = 1/11$ for all $\omega \in \Omega$. Calculate and plot the expected utility $U(\xi, a) = \sum_\omega \xi(\omega) U(\omega, a)$ for each a. Report $\max_a U(\xi, a)$.

Exercise 3.7.2 Assume $\omega \in \Omega$ is arbitrary (but deterministically selected). Calculate the utility $U(a) = \min_\omega U(\omega, a)$ for each a. Report $\max(U)$.

Exercise 3.7.3 Again assume $\omega \in \Omega$ is arbitrary (but deterministically selected). We now allow for stochastic policies π on $\mathcal{A}$. Then the expected utility is $U(\omega, \pi) = \sum_a U(\omega, a)\pi(a)$.

(a) Calculate and plot the expected utility when $\pi(a) = 1/11$ for all a, reporting values for all ω.
(b) Find

$$\max_\pi \min_\xi U(\xi, \pi).$$

Hint: Use the linear programming formulation, adding a constant to the utility matrix U so that all elements are non-negative.

Exercise 3.7.4 Consider the definition of rules that, for some $\epsilon > 0$, select a maximizing

$$P\big(\{\omega \mid U(\omega, a) > \sup_{d' \in \mathcal{A}} U(\omega, d') - \epsilon\}\big).$$

Prove that this is indeed a statistical decision problem, i.e., it corresponds to maximizing the expectation of some utility function.

3.7.2 Problems with Observations

For the following exercises we consider a set of worlds Ω and a decision set $\mathcal{A}$, as well as the following utility function $U : \Omega \times \mathcal{A} \to \mathbb{R}$:

$$U(\omega, a) = -|\omega - a|^2$$

In addition, we consider a family of distributions on a sample space $S = \{0, 1\}^n$,

$$\mathscr{F} \triangleq f_\omega \omega \in \Omega,$$

such that f_ω is the binomial probability mass function with parameters ω. Consider the parameter set

$$\Omega = \{0, 0.1, \dots, 0.9, 1\}.$$

Let ξ be the uniform distribution on Ω, such that $\xi(\omega) = 1/11$ for all $\omega \in \Omega$. Further, let the decision set be $\mathcal{A} = [0, 1]$.

Exercise 3.7.5 What is the decision a^* maximizing $U(\xi, a) = \sum_\omega \xi(\omega) U(\omega, a)$ and what is $U(\xi, a^*)$?

Exercise 3.7.6 In the same setting, we now observe the sequence $x = (x_1, x_2, x_3) = (1, 0, 1)$.

1. Plot the posterior distribution $\xi(\omega \mid x)$ and compare it to the posterior we would obtain if our prior on ω was $\xi' = \mathcal{Beta}(2, 2)$.
2. Find the decision a^* maximizing the *a posteriori* expected utility

$$\mathbb{E}_\xi(U \mid a, x) = \sum_\omega U(\omega, a) \xi(\omega \mid x).$$

3. Consider $n = 2$, i.e., $S = \{0, 1\}^2$. Calculate the Bayes-optimal expected utility in extensive form:

$$\mathbb{E}_\xi(U \mid \pi^*) = \sum_S \phi(x) \sum_\omega U[\omega, \pi^*(x)] \xi(\omega \mid x) = \sum_S \phi(x) \max_a \sum_\omega U[\omega, a] \xi(\omega \mid x),$$

where $\phi(x) = \sum_\omega f_\omega(x) \xi(\omega)$ is the prior marginal distribution of x and $\delta^* : S \to \mathcal{A}$ is the Bayes-optimal decision rule.
Hint: You can simplify the computational complexity somewhat, since you only need to calculate the probability of $\sum_t x_t$. This is not necessary to solve the problem though.

Exercise 3.7.7 In the same setting, we consider nature to be adversarial. Once more, we observe $x = (1, 0, 1)$. Assume that nature can choose a prior among a set of priors $\Xi = \{\xi_1, \xi_2\}$. Let $\xi_1(\omega) = 1/11$ and $\xi_2(\omega) = \omega/5.5$ for each ω.

1. Calculate and plot the value for deterministic decisions a:

$$\min_{\xi \in \Xi} \mathbb{E}_\xi(U \mid a, x).$$

2. Find the minimax prior ξ^*

$$\min_{\xi \in \Xi} \max_{a \in \mathcal{A}} \mathbb{E}_\xi(U \mid a).$$

Hint: Apart from the adversarial prior selection, this is very similar to the previous exercise.

3.7.3 An Insurance Problem

Consider Example 2.15 of Chap. 2. Therein, an insurer is covering customers by asking for a premium $d > 0$. In the event of an accident, which happens with probability $\epsilon \in [0, 1]$, the insurer pays out $h > 0$. The problem of the insurer is, given ϵ, h, what to set the premium d to so as to maximize its expected utility. We assume that the insurer's utility is linear.

We now consider customers with some baseline income level $x \in \mathcal{S}$. For simplicity, we assume that the only possible income levels are in $\mathcal{S} = \{15, 20, 25, \ldots, 60\}$, with $K = 10$ possible incomes. Let $V : \mathbb{R} \to \mathbb{R}$ denote the utility function of a customer. Customers who are *interested* in the insurance product will buy if and only if

$$V(x - d) > \epsilon V(x - h) + (1 - \epsilon) V(x).$$

We make the simplifying assumption that the utility function is the same for all customers and has the following form:

$$V(x) = \begin{cases} \ln x, & x \geq 1 \\ 1 - (x - 2)^2, & \text{otherwise.} \end{cases}$$

Customers who are *not* interested the insurance product, will not buy it no matter what the price.

There is some unknown probability distribution $P_\omega(x)$ over the income level, such that the probability of n people having incomes $x^n = (x_1, \ldots, x_n)$ is $P^n_\omega(x^n) = \prod_{i=1}^n P_\omega(x_i)$. We have two data sources for this. The first is a model of the general population ω_1 not working in high-tec industry, and the second is a model of employees in high-tech industry, ω_2. The models are summarized in Table 3.6 below. Together, these two models form a family of distributions $\mathscr{P} = \{P_\omega \mid \omega \in \Omega\}$ with $\Omega = \{\omega_1, \omega_2\}$.

Table 3.6 Income level distribution for the two models

Income levels	15	20	25	30	35	40	45	50	55	60
Models	Probability (%) of income level $P_\omega(x)$									
ω_1	5	10	12	13	11	10	8	10	11	10
ω_2	8	4	1	6	11	14	16	15	13	12

The goal is to find a premium d that maximizes the expected utility of the insurance company. We assume that the company is liquid enough that utility is linear. In the following, we consider four different cases. For simplicity, you can let $\mathcal{A} = \mathcal{S}$ throughout this exercise.

Exercise 3.7.8 Show that the expected utility for a given ω is the expected gain from a buying customer times the probability that an interested customer will have an income x such that she would buy our insurance, that is,

$$U(\omega, d) = (d - \epsilon h) \sum_{x \in \mathcal{S}} P_\omega(x) \mathbb{I} \{V(x - d) > \epsilon V(x - h) + (1 - \epsilon) V(x)\} .$$

Let $h = 150$ and $\epsilon = 10^{-3}$. Plot the expected utility for the two possible ω for varying d. What is the optimal price level if the incomes of all interested customers are distributed according to ω_1? What is the optimal price level if they are distributed according to ω_2?

Exercise 3.7.9 According to our intuition, customers interested in our product are much more likely to come from the high-tech industry than from the general population. For that reason, we have a prior probability $\xi(\omega_1) = 1/4$ and $\xi(\omega_2) = 3/4$ over the parameters Ω of the family $\mathscr{P}$. More specifically, and in keeping with our previous assumptions, we formulate the following model:

$$\omega^* \sim \xi$$
$$x^T \mid \omega^* = \omega \sim P_\omega$$

That is, the data is drawn from one unknown model $\omega^* \in \Omega$. This can be thought of as an experiment where nature randomly selects ω^* with probability ξ and then generates the data from the corresponding model P_{ω^*}. Plot the expected utility under this prior as the premium d varies. What is the optimal expected utility and premium?

Exercise 3.7.10 Instead of fully relying on our prior, the company decides to perform a random survey of 1000 people who are asked whether they would be interested in the insurance product (as long as the price is low enough). If interested, they are also asked about their income level. Assume that only 126 people were interested, with income levels as given in Table 3.7. Each row column of the table shows the stated income and the number of people reporting it.

Table 3.7 Survey results

Income	15	20	25	30	35	40	45	50	55	60
Number	7	8	7	10	15	16	13	19	17	14

Let $x^n = \{x_1, x_2, \dots, x_n\}$ be the set of data collected. Assuming that that the responses are truthful, calculate the posterior probability $\xi(\omega \mid x^n)$, assuming that the only possible models of income distribution are the two models ω_1, ω_2 used in the previous exercises. Plot the expected utility under the posterior distribution as d varies. What is the maximum expected utility that can be obtained?

Exercise 3.7.11 Having only two possible models is somewhat limiting, especially since neither of them might correspond to the income distribution of people interested in our insurance product. How could this problem be rectified? Describe your idea and implement it. When would you expect this to work better?

3.7.4 *Medical Diagnosis*

Many patients arriving at an emergency room suffer from chest pain. This may indicate acute coronary syndrome (ACS). Patients suffering from ACS that go untreated may die with probability[5] 2% in the next few days. Successful diagnosis lowers the short-term mortality rate to 0.2%. Consequently, a prompt diagnosis is essential.

Statistics of patients Approximately 50% of patients presenting with chest pain turn out to suffer from ACS (either acute myocardial infraction or unstable angina pectoris). Approximately 10% suffer from lung cancer. Of ACS sufferers in general, $2/3$ are smokers and $1/3$ non-smokers. Only $1/4$ of non-ACS sufferers are smokers. In addition, 90% of lung cancer patients are smokers. Only $1/4$ of non-cancer patients are smokers.

Assumption 3.7.1 A patient may suffer from none, either or both conditions.

Assumption 3.7.2 When the smoking history of the patient is known, the development of cancer or ACS are independent.

Tests One can perform an ECG to test for ACS. An ECG test has *sensitivity* of 66.6% (i.e., it correctly detects $2/3$ of all patients that suffer from ACS), and a *specificity* of 75% (i.e., $1/4$ of patients that do not have ACS, still test positive). An X-ray can diagnose lung cancer with a sensitivity of 90% and a specificity of 90%.

Assumption 3.7.3 Repeated applications of a test produce the same result for the same patient, i.e., that randomness is only due to patient variability.

[5] The following figures are not really accurate, as they are liberally adapted from different studies.

Assumption 3.7.4 The existence of lung cancer *does not* affect the probability that the ECG will be positive. Conversely, the existence of ACS *does not* affect the probability that the X-ray will be positive.

The main problem we want to solve, is how to perform experiments or tests, so as to diagnose the patient using as few resources as possible and making sure the patient lives. This is a problem in *experiment design*. We start from the simplest case and look at a couple of examples where we only observe the results of some tests. We then examine the case where we can select which tests to perform.

Exercise 3.7.12 Now consider the case where you have the choice which tests to perform. First, you observe S, i.e., whether or not the patient is a smoker. Then you select a test $d_1 \in \{\texttt{X-ray}, \texttt{ECG}\}$ to make. Finally, you decide whether or not to treat for ASC, that is, you choose $d_2 \in \{\texttt{heart treatment}, \texttt{no treatment}\}$. An untreated ASC patient may die with probability 2%, while a treated one with probability 0.2%. Treating a non-ASC patient results in death with probability 0.1%.

1. Draw a decision diagram, where:
   - S is an observed random variable taking values in $\{0, 1\}$.
   - A is a hidden variable taking values in $\{0, 1\}$.
   - C is a hidden variable taking values in $\{0, 1\}$.
   - d_1 is a choice variable, taking values in $\{\texttt{X-ray}, \texttt{ECG}\}$.
   - r_1 is a result variable, taking values in $\{0, 1\}$, corresponding to negative and positive tests results.
   - d_2 is a choice variable, which depends on the test result of d_1 and on S.
   - r_2 is a result variable, taking values in $\{0, 1\}$ corresponding to the patient dying (0), or living (1).
2. Let $d_1 = \texttt{X-ray}$, and assume the patient suffers from ACS, i.e., $A = 1$. How is the posterior distributed?
3. What is the optimal decision rule for this problem?

References

1. Leonard, J.: Savage. The Foundations of Statistics. Dover Publications (1972)
2. Kearns, M., Roth, A.: The Ethical Algorithm: The Science of Socially Aware Algorithm Design. Oxford University Press, USA (2019)
3. Dwork, C., Hardt, M., Pitassi, T., Reingold, O., Zemel, R.: Fairness through awareness. In: Proceedings of the 3rd Innovations in Theoretical Computer Science Conference, pp. 214–226. ACM (2012)
4. Chouldechova, A.: Fair prediction with disparate impact: a study of bias in recidivism prediction instruments. Technical Report (2016). arXiv:1610.07524
5. Corbett-Davies, S., Pierson, E., Feller, A., Goel, S., Huq, A.: Algorithmic decision making and the cost of fairness. Technical Report (2017). arXiv:1701.08230

6. Kleinberg, J., Mullainathan, S., Raghavan, M.: Inherent trade-offs in the fair determination of risk scores. Technical Report (2016). arXiv:1609.05807
7. Kilbertus, N., Rojas-Carulla, M., Parascandolo, G., Hardt, M., Janzing, D., Schölkopf, B.: Avoiding discrimination through causal reasoning. Technical Report (2017). arXiv:1706.02744
8. Dimitrakakis, C., Liu, Y., Parkes, D., Radanovic, G.: Subjective fairness: fairness is in the eye of the beholder. Technical Report (2017). arXiv:1706.00119

Chapter 4
Estimation

4.1 Introduction

In the previous chapter, we have seen how to make optimal decisions with respect to a given utility function and belief. One important question is how to compute an updated belief from observations and a prior belief. More generally, we wish to examine how much information we can obtain about an unknown parameter from observations, and how to bound the respective estimation error. While most of this chapter will focus on the Bayesian framework for estimating parameters, we shall also look at tools for making conclusions about the value of parameters without making specific assumptions about the data distribution, i.e., without providing specific prior information.

In the Bayesian setting, we calculate posterior distributions of parameters given data. The basic problem can be stated as follows. Let $\mathscr{P} \triangleq \{P_\omega \mid \omega \in \Omega\}$ be a family of probability measures on $(\mathcal{S}, \mathcal{F}_\mathcal{S})$ and ξ be our prior probability measure on $(\Omega, \mathcal{F}_\Omega)$. Given some data $x \sim P_{\omega^*}$, with $\omega^* \in \Omega$, how can we estimate ω^*? The Bayesian approach is to estimate the posterior distribution $\xi(\cdot \mid x)$, instead of guessing a single ω^*. In general, the posterior measure is a function $\xi(\cdot \mid x) : \mathcal{F}_\Omega \to [0, 1]$, with

$$\xi(B \mid x) = \frac{\int_B P_\omega(x)\, \mathrm{d}\xi(\omega)}{\int_\Omega P_\omega(x)\, \mathrm{d}\xi(\omega)}. \tag{4.1.1}$$

The posterior distribution allows us to quantify our uncertainty about the unknown ω^*. This in turn enables us to take decisions that take uncertainty into account.

The first question we are concerned with in this chapter is how to calculate this posterior for any value of x in practice. If x is a complex object, this may be computationally difficult. In fact, the posterior distribution can also be a complicated function. However, there exist distribution families and priors such that this calculation is very easy, in the sense that the functional form of the posterior depends upon a

C. Dimitrakakis and R. Ortner, *Decision Making Under Uncertainty and Reinforcement Learning*, Intelligent Systems Reference Library 223,
https://doi.org/10.1007/978-3-031-07614-5_4

small number of parameters. This happens when a summary of the data that contains all necessary information can be calculated easily. Formally, this is captured via the concept of a *sufficient statistic*.

4.2 Sufficient Statistics

Sometimes we want to summarize the data we have observed. This can happen when the data is a long sequence of simple observations $x^n = (x_1, \dots, x_n)$. It may also be useful to do so when we have a single observation x, such as a high-resolution image. For some applications, it may be sufficient to only calculate a really simple function of the data, such as the sample mean.

Definition 4.2.1 (*Sample mean*) The sample mean $\bar{x}_n : \mathbb{R}^n \to \mathbb{R}$ of a sequence $x^n = (x_1, \dots, x_n)$ with $x_t \in \mathbb{R}$ is defined as

$$\bar{x}_n \triangleq \frac{1}{n} \sum_{t=1}^{n} x_t .$$

statistic

The mean is an example for what is called a *statistic*, that is, a function of the observations to some vector space. In the following, we are interested in statistics that can replace all the complete original data in our calculations without losing any information. Such statistics are called *sufficient*.

4.2.1 *Sufficient Statistics*

We consider the standard probabilistic setting. Let $\mathcal{S}$ be a sample space and Ω be a parameter space defining a family of measures on $\mathcal{S}$, that is,

$$\mathscr{P} = \{P_\omega \mid \omega \in \Omega\} .$$

In addition, we must also define an appropriate prior distribution ξ on the parameter space Ω. Then the definition of a sufficient statistic in the Bayesian sense[1] is as follows.

Definition 4.2.2 Let Ξ be a set of prior distributions on Ω, which indexes a family $\mathscr{P} = \{P_\omega \mid \omega \in \Omega\}$ of distributions on $\mathcal{S}$. A statistic $\phi : \mathcal{S} \to \mathcal{Z}$, where $\mathcal{Z}$ is a vector space,[2] is a *sufficient statistic* for $\langle \mathscr{P}, \Xi \rangle$, if

---

[1] There is an alternative definition, which replaces equality of posterior distributions with point-wise equality on the family members, i.e., $P_\omega(x) = P_\omega(x')$ for all ω. This is a stronger definition, as it implies the Bayesian one we use here.

[2] Typically $\mathcal{Z} \subset \mathbb{R}^k$ for finite-dimensional statistics.

$$\xi(\cdot \mid x) = \xi(\cdot \mid x') \tag{4.2.1}$$

for any prior $\xi \in \Xi$ and any $x, x' \in \mathcal{S}$ such that $\phi(x) = \phi(x')$.

This means that the statistic is sufficient if, whenever we obtain the same value of the statistic for two different datasets x, x', then the resulting posterior distribution over the parameters is identical, independent of the prior distribution. In other words, the value of the statistic is sufficient for computing the posterior. Interestingly, a sufficient statistic always implies the following factorization for members of the family.

Theorem 4.2.1 *A statistic $\phi : \mathcal{S} \to \mathcal{Z}$ is sufficient for $\langle \mathscr{P}, \Xi \rangle$ iff there exist functions $u : \mathcal{S} \to (0, \infty)$ and $v : \mathcal{Z} \times \Omega \to [0, \infty)$ such that $\forall x \in \mathcal{S}, \omega \in \Omega$:*

$$P_\omega(x) = u(x)\, v[\phi(x), \omega]$$

Proof In the following proof we assume arbitrary Ω. The case when Ω is finite is technically simpler and is left as an exercise. Let us first assume the existence of u, v satisfying the equation. Then for any $B \in \mathcal{F}_\Omega$ we have

$$\begin{aligned}\xi(B \mid x) &= \frac{\int_B u(x)\, v[\phi(x), \omega]\, \mathrm{d}\xi(\omega)}{\int_\Omega u(x)\, v[\phi(x), \omega]\, \mathrm{d}\xi(\omega)} \\ &= \frac{\int_B v[\phi(x), \omega]\, \mathrm{d}\xi(\omega)}{\int_\Omega v[\phi(x), \omega]\, \mathrm{d}\xi(\omega)}\end{aligned}$$

For x' with $\phi(x) = \phi(x')$, it follows that $\xi(B \mid x) = \xi(B \mid x')$, so $\xi(\cdot \mid x) = \xi(\cdot \mid x')$ and ϕ satisfies the definition of a sufficient statistic.

Conversely, assume that ϕ is a sufficient statistic. Let μ be a dominating measure on $\mathcal{S}$ so that we can define the densities $p(\omega) \triangleq \frac{\mathrm{d}\xi(\omega)}{\mathrm{d}\mu(\omega)}$ and

$$p(\omega \mid x) \triangleq \frac{\mathrm{d}\xi(\omega \mid x)}{\mathrm{d}\mu(\omega)} = \frac{P_\omega(x)\, p(\omega)}{\int_\Omega P_\omega(x)\, \mathrm{d}\xi(\omega)},$$

whence

$$P_\omega(x) = \frac{p(\omega \mid x)}{p(\omega)} \int_\Omega P_\omega(x)\, \mathrm{d}\xi(\omega).$$

Since ϕ is sufficient, there is by definition some function $g : \mathcal{Z} \times \Omega \to [0, \infty)$ such that $p(\omega \mid x) = g[\phi(x), \omega]$. This must be the case, since sufficiency means that $p(\omega \mid x) = p(\omega \mid x')$ whenever $\phi(x) = \phi(x')$. For this to occur the existence of such a g is necessary. Consequently, we can factorize P_ω as

$$P_\omega(x) = u(x)\, v[\phi(x), \omega],$$

where $u(x) = \int_\Omega P_\omega(x)\, \mathrm{d}\xi(\omega)$ and $v[\phi(x), \omega] = g[\phi(x), \omega]/p(\omega)$. □

In the factorization of Theorem 4.2.1, u is the only factor that depends directly on x. Interestingly, it *does not appear* in the posterior calculation at all. So, the posterior only depends on x through the statistic.

Example 4.1 Suppose $x^n = (x_1, \ldots, x_n)$ is a random sample from a distribution with two possible outcomes 0 and 1 where ω is the probability of outcome 1. (This is usually known as a Bernoulli distribution that we will have a closer look at in Sect. 4.3.1). Then the probability of observing a certain sequence x^n is given by

$$P_\omega(x^n) = \prod_{t=1}^{n} P_\omega(x_t) = \omega^{s_n}(1-\omega)^{n-s_n},$$

where $s_n = \sum_{t=1}^n x_t$ is the number of times 1 has been observed until time n. Here the statistic $\phi(x^n) = s_n$ satisfies (4.2.1) with $u(x) = 1$, while $P_\omega(x^n)$ only depends on the data through the statistic $s_n = \phi(x^n)$.

Another example is when we have a *finite* set of models. Then the sufficient statistic is always a finite-dimensional vector.

Lemma 4.2.1 *Let $\mathscr{P} = \{P_\theta \mid \theta \in \Theta\}$ be a family, where each model P_θ is a probability measure on $\mathcal{X}$ and Θ contains n models. If $\mathbf{p} \in \mathbb{\Delta}^n$ is a vector representing our prior distribution, i.e., $\xi(\theta) = p_\theta$, then the finite-dimensional vector with entries $q_\theta = p_\theta P_\theta(x)$ is a sufficient statistic.*

Proof Simply note that the posterior distribution in this case is

$$\xi(\theta \mid x) = \frac{q_\theta}{\sum_{\theta'} q_{\theta'}},$$

that is, the values q_θ are sufficient to compute the posterior. □

From the proof it is clear that also the vector with entries $w_\theta = \frac{q_\theta}{\sum_{\theta'} q_{\theta'}}$ is a sufficient statistic.

4.2.2 Exponential Families

Even when dealing with an infinite set of models in some cases the posterior distributions can be computed efficiently. Many well-known distributions such as the Gaussian, Bernoulli and Dirichlet distribution are members of the exponential family of distributions. All those distributions are factorizable in the manner shown below, while at the same time they have fixed-dimension sufficient statistics.

Definition 4.2.3 A distribution family $\mathscr{P} = \{P_\omega \mid \omega \in \Omega\}$ with P_ω being a probability function (or density) on the sample space $\mathcal{S}$ is said to be an *exponential family* if there are suitable functions $a : \Omega \to \mathbb{R}$, $b : \mathcal{S} \to \mathbb{R}$, $g_i : \Omega \to \mathbb{R}$ and $h_i : \mathcal{S} \to \mathbb{R}$ $(1 \leq i \leq k)$ such that for any $x \in \mathcal{S}$, $\omega \in \Omega$:

$$P_\omega(x) = a(\omega)\, b(x) \exp\left[\sum_{i=1}^{k} g_i(\omega)\, h_i(x)\right]$$

Among families of distributions satisfying certain smoothness conditions, only exponential familes have a fixed-dimension sufficient statistic. Because of this, exponential family distributions admit so-called parametric *conjugate* prior distribution families. These have the property that any posterior distribution calculated will remain within the conjugate family. Frequently, because of the simplicity of the statistic used, calculation of the conjugate posterior parameters is very simple.

4.3 Conjugate Priors

In this section, we examine some well-known conjugate families. First, we give sufficient conditions for the of existence of a conjugate family of priors for a given distribution family and statistic. While this section can be used as a reference, the reader may wish to initially only look at the first few example families.

The following remark gives sufficient conditions for the existence of a finite-dimensional sufficient statistic.

Remark 4.3.1 If a family $\mathscr{P}$ of distributions on $\mathcal{S}$ has a sufficient statistic $\phi : \mathcal{S} \to \mathcal{Z}$ of *fixed* dimension for any $x \in \mathcal{S}$, then there exists a conjugate family of priors $\Xi = \{\xi_\alpha \mid \alpha \in A\}$ with a set A of possible parameters for the prior distribution, such that:

1. $P_\omega(x)$ is proportional to some $\xi_\alpha \in \Xi$, that is,

$$\forall x \in \mathcal{S}\ \exists \xi_\alpha \in \Xi, c > 0 : \int_B P_\omega(x)\, \mathrm{d}\xi_\alpha(\omega) = c\, \xi_\alpha(B), \forall B \in \mathcal{F}_\Omega.$$

2. The family is closed under multiplication, that is,

$$\forall \xi_1, \xi_2 \in \Xi\ \exists \xi_\alpha \in \Xi, c > 0 : \xi_\alpha = c\, \xi_1 \xi_2.$$

While conjugate families exist for statistics with unbounded dimension, here we shall focus on finite-dimensional families. We will start with the simplest example, the Bernoulli-Beta pair.

4.3.1 Bernoulli-Beta Conjugate Pair

The Bernoulli distribution is a discrete distribution that is ideal for modelling independent random trials with just two different outcomes (typically 'success' and 'failure') and a fixed probability of success.

Definition 4.3.1 (*Bernoulli distribution*) The Bernoulli distribution is discrete with outcomes $\mathcal{S} = \{0, 1\}$, a parameter $\omega \in [0, 1]$, and probability function

$$P_\omega(x = u) = \omega^u (1 - \omega)^{1-u} = \begin{cases} \omega, & \text{if } u = 1 \\ 1 - \omega, & \text{if } u = 0. \end{cases}$$

If x is distributed according to a Bernoulli distribution with parameter ω, we write $x \sim \mathcal{Bern}(\omega)$.

The structure of the graphical model in Fig. 4.1 shows the dependencies between the different variables of the model.

When considering n independent trials of a Bernoulli distribution the set of possible outcomes is $\mathcal{S} = \{0, 1\}^n$. Then $P_\omega(x^n) = \prod_{t=1}^n P_\omega(x_t)$ is the probability of observing the exact sequence x^n under the Bernoulli model. However, in many cases we are interested in the probability of observing a particular number of successes (outcome 1) and failures (outcome 0) and do not care about the actual order. For summarizing we need to *count* the actual sequences in which out of n trials we have k successes. The actual number is given by the *binomial coefficient*, defined as

binomial coefficient

$$\binom{n}{k} = \frac{n!}{k!\,(n-k)!} \qquad k, n \in \mathbb{N}, n \geq k.$$

Now we are ready to define the binomial distribution, that is a scaled product-Bernoulli distribution for multiple independent outcomes where we want to measure the probability of a particular number successes or failures. Thus, the Bernoulli is a distribution on a sequence of outcomes, while the binomial is a distribution on the total number of successes. That is, let $s = \sum_{t=1}^n x_t$ be the total number of successes observed up to time n. Then we are interested in the probability that there are exactly k successes out of n trials.

Definition 4.3.2 (*Binomial distribution*) The binomial distribution with parameters ω and n has outcomes $\mathcal{S} = \{0, 1, \ldots, n\}$. Its probability function is given by

$$P_\omega(s = k) = \binom{n}{k} \omega^k (1 - \omega)^{n-k}.$$

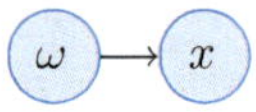

Fig. 4.1 Bernoulli graphical model

If s is drawn from a binomial distribution with parameters ω, n, we write $s \sim \mathcal{B}inom(\omega, n)$.

Now let us return to the Bernoulli distribution. If the parameter ω is known, then observations are independent of each other. However, this is not the case when ω is unknown. For example, if $\Omega = \{\omega_1, \omega_2\}$, then the probability of observing a sequence x^n is given by

$$\mathbb{P}(x^n) = \sum_{\omega \in \Omega} \mathbb{P}(x^n \mid \omega)\mathbb{P}(\omega) = \prod_{t=1}^{n} \mathbb{P}(x_t \mid \omega_1)\mathbb{P}(\omega_1) + \prod_{t=1}^{n} \mathbb{P}(x_t \mid \omega_2)\mathbb{P}(\omega_2),$$

which in general is different from $\prod_{t=1}^{n} \mathbb{P}(x_t)$. For the general case where $\Omega = [0, 1]$ the question is whether there is a prior distribution that can succinctly describe our uncertainty about the parameter. Indeed, there is, and it is called the *Beta distribution*. It is defined on the interval $[0, 1]$ and has two parameters that determine the density of the observations. Because the Bernoulli distribution has a parameter in $[0, 1]$, the outcomes of the Beta can be used to specify a prior on the parameters of the Bernoulli distribution.

Beta distribution

Definition 4.3.3 (*Beta distribution*) The Beta distribution has outcomes $\omega \in \Omega = [0, 1]$ and parameters $\alpha_0, \alpha_1 > 0$, which we will summarize in a vector $\boldsymbol{\alpha} = (\alpha_1, \alpha_0)$. Its probability density function is given by

$$p(\omega \mid \boldsymbol{\alpha}) = \frac{\Gamma(\alpha_0 + \alpha_1)}{\Gamma(\alpha_0)\Gamma(\alpha_1)} \omega^{\alpha_1 - 1}(1 - \omega)^{\alpha_0 - 1}, \tag{4.3.1}$$

where $\Gamma(\alpha) = \int_0^\infty u^{\alpha-1} e^{-u} \, du$ is the *Gamma function*. If ω is distributed according to a Beta distribution with parameters α_1, α_0, we write $\omega \sim \mathcal{B}eta(\alpha_1, \alpha_0)$.

Gamma function

We note that the Gamma function is an extension of the function $n!$, so that for $n \in \mathbb{N}$ it holds that $\Gamma(n) = n!$. That way, the first term in (4.3.1) corresponds to a generalized binomial coefficient. The dependencies between the parameters are shown in the graphical model of Fig. 4.2. A Beta distribution with parameter $\boldsymbol{\alpha}$ has expectation

$$\mathbb{E}(\omega \mid \boldsymbol{\alpha}) = \frac{\alpha_1}{\alpha_0 + \alpha_1}$$

and variance

$$\mathbb{V}(\omega \mid \boldsymbol{\alpha}) = \frac{\alpha_1 \alpha_0}{(\alpha_1 + \alpha_0)^2(\alpha_1 + \alpha_0 + 1)}.$$

Fig. 4.2 Beta graphical model

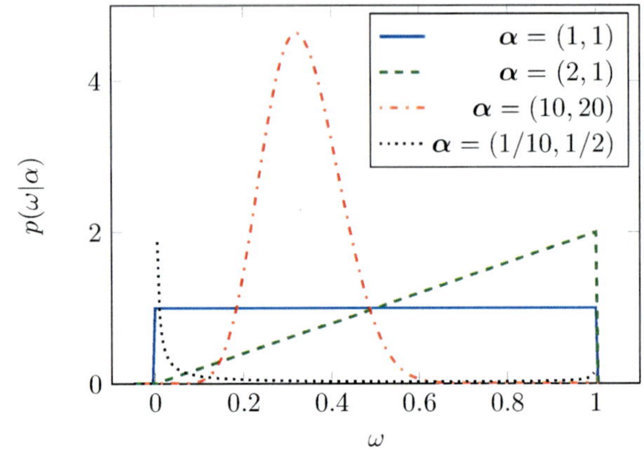

Fig. 4.3 Four example Beta densities

Figure 4.3 shows the density of a Beta distribution for four different parameter vectors. When $\alpha_0 = \alpha_1 = 1$, the distribution is equivalent to a uniform one.

4.3.1.1 Beta Prior for Bernoulli Distributions

As already indicated, we can encode our uncertainty about the unknown but fixed parameter $\omega \in [0, 1]$ of a Bernoulli distribution using a Beta distribution. We start with a Beta distribution with parameter $\boldsymbol{\alpha}$ that defines our prior ξ_0 via $\xi_0(B) \triangleq \int_B p(\omega \mid \boldsymbol{\alpha})\, d\omega$. Then the posterior probability is given by

$$p(\omega \mid x^n, \boldsymbol{\alpha}) = \frac{\prod_{t=1}^n P_\omega(x_t)\, p(\omega \mid \boldsymbol{\alpha})}{\int_\Omega \prod_{t=1}^n P_\omega(x_t)\, p(\omega \mid \boldsymbol{\alpha})\, d\omega} \propto \omega^{s_{n,1}+\alpha_1-1}(1-\omega)^{s_{n,0}+\alpha_0-1},$$

where $s_{n,1} = \sum_{t=1}^n x_t$ and $s_{n,0} = n - s_{n,1}$ are the total number of 1s and 0s, respectively. As can be seen, this again has the form of a Beta distribution (Fig. 4.4).

Beta-Bernoulli model

Let ω be drawn from a Beta distribution with parameters α_1, α_0, and $x^n = (x_1, \ldots, x_n)$ be a sample drawn independently from a Bernoulli distribution with parameter ω, i.e.,

$$\omega \sim \mathcal{B}eta(\alpha_1, \alpha_0), \qquad x^n \mid \omega \sim \mathcal{B}ern^n(\omega).$$

Fig. 4.4 Beta-Bernoulli graphical model

Then the posterior distribution of ω given the sample the posterior distribution is also Beta, that is,

$$\omega \mid x^n \sim \mathcal{B}eta(\alpha_1', \alpha_0') \quad \text{with} \quad \alpha_1' = \alpha_1 + s_{n,1}, \quad \alpha_0' = \alpha_0 + s_{n,0}.$$

Example 4.2 The parameter $\omega \in [0, 1]$ of a randomly selected coin can be modelled as a Beta distribution peaking around $\frac{1}{2}$. Usually one assumes that coins are fair. However, not all coins are exactly the same. Thus, it is possible that each coin deviates slightly from fairness. We can use a Beta distribution to model how likely (we think) different values ω of coin parameters are.

To demonstrate how belief changes, we perform the following simple experiment. We repeatedly toss a coin and wish to form an accurate belief about how biased the coin is, under the assumption that the outcomes are Bernoulli with a fixed parameter ω. Our initial belief, ξ_0, is modelled as a Beta distribution on the parameter space $\Omega = [0, 1]$, with parameters $\alpha_0 = \alpha_1 = 100$. This places a strong prior on the coin being close to fair. However, we still allow for the possibility that the coin is biased.

Figure 4.5 shows a sequence of beliefs at times 0, 10, 100, 1000 respectively, from a coin with bias $\omega = 0.6$. Due to the strength of our prior, after 10 observations, the situation has not changed much and the belief ξ_{10} is very close to the initial one. However, after 100 observations our belief has now shifted towards 0.6, the true bias of the coin. After a total of 1000 observations, our belief is centered very close to 0.6, and is now much more concentrated, reflecting the fact that we are almost certain about the value of ω.

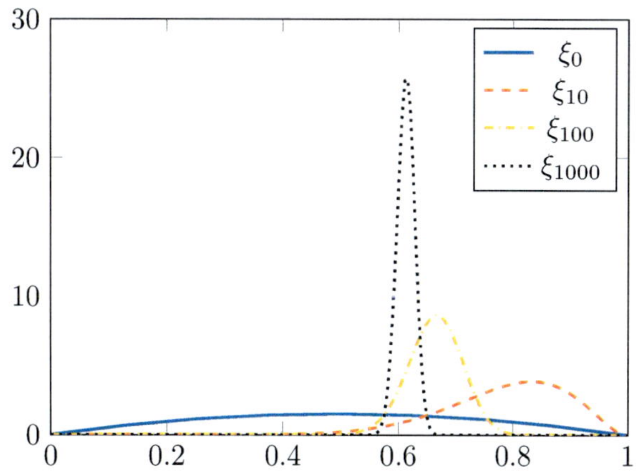

Fig. 4.5 Changing beliefs as we observe tosses from a coin with probability $\omega = 0.6$ of heads

4.3.2 Conjugates for the Normal Distribution

The well-known normal distribution is also endowed with suitable conjugate priors. We first give the definition of the normal distribution, then consider the cases where we wish to estimate its mean, its variance, or both at the same time.

Definition 4.3.4 (*Normal distribution*) The normal distribution is a continuous distribution with outcomes in $\mathbb{R}$. It has two parameters, the mean $\omega \in \mathbb{R}$ and the variance $\sigma^2 \in \mathbb{R}^+$, or alternatively the precision $r \in \mathbb{R}^+$, where $\sigma^2 = r^{-1}$. Its probability density function is given by

$$f(x \mid \omega, r) = \sqrt{\frac{r}{2\pi}} \exp\left(-\frac{r}{2}(x-\omega)^2\right).$$

When x is distributed according to a normal distribution with parameters ω, r^{-1}, we write $x \sim \mathcal{N}(\omega, r^{-1})$.

For a respective sample x^n of size n, we write $x^n \sim \mathcal{N}^t(\omega, r^{-1})$. Independent samples satisfy the independence condition

$$f(x^n \mid \omega, r) = \prod_{t=1}^{n} f(x_t \mid \omega, r) = \left(\sqrt{\frac{r}{2\pi}}\right)^n \exp\left(-\frac{r}{2}\sum_{t=1}^{n}(x_t - \omega)^2\right).$$

The dependency graph in Fig. 4.6 shows the dependencies between the parameters of a normal distribution and the observations x_t. In this graph, only a single sample x_t is shown, and it is implied that all x_t are independent of each other given r, ω.

Transformations of normal samples. The importance of the normal distribution stems from the fact that many actual distributions turn out to be approximately normal. Another interesting property of the normal distribution concerns transformations of normal samples. For example, if x^n is drawn from a normal distribution with mean ω and precision r, then $\sum_{t=1}^n x_t \sim \mathcal{N}(n\omega, nr^{-1})$. Finally, if the samples x_t are drawn from the *standard normal* distribution, i.e., $x_t \sim \mathcal{N}(0, 1)$, then $\sum_{t=1}^n x_t^2$ has a χ^2-*distribution* with n degrees of freedom (cf. also the discussion of the Gamma distribution below).

standard normal

χ^2-distribution

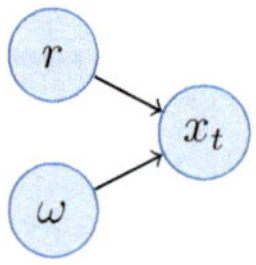

Fig. 4.6 Normal graphical model

4.3.2.1 Normal Distribution with Known Precision, Unknown Mean

The simplest normal estimation problem occurs when we only need to estimate the mean and assume that the variance (or equivalently, the precision) is known. For Bayesian estimation, it is convenient to assume that the mean ω is drawn from *another* normal distribution with known mean. This gives a conjugate pair and thus results in a posterior normal distribution for the mean as well (Fig. 4.7).

Normal-Normal conjugate pair
Let ω be drawn from a normal distribution with mean μ and precision τ, and $x^n = (x_1, \ldots, x_n)$ be a sample drawn independently from a normal distribution with mean ω and precision r, that is,

$$x^n \mid \omega, r \sim \mathcal{N}^n(\omega, r^{-1}), \qquad \omega \mid \tau \sim \mathcal{N}(\mu, \tau^{-1}).$$

Then the posterior distribution of ω given the sample is also normal, that is,

$$\omega \mid x^n \sim \mathcal{N}(\mu', \tau'^{-1}) \qquad \text{with } \mu' = \frac{\tau\mu + nr\bar{x}_n}{\tau'}, \qquad \tau' = \tau + nr,$$

where $\bar{x}_n \triangleq \frac{1}{n}\sum_{t=1}^{n} x_t$.

It can be seen that the updated estimate for the mean is shifted towards the empirical mean $\bar{x}_n$, and the precision increases linearly with the number of samples.

4.3.2.2 Normal with Unknown Precision and Known Mean

To model normal distributions with known mean, but unknown precision (or equivalently, unknown variance), we first have to introduce the Gamma distribution that we will use to represent our uncertainty about the precision.

Definition 4.3.5 (*Gamma distribution*) The Gamma distribution is a continuous distribution with outcomes in $[0, \infty)$. It has two parameters $\alpha > 0$, $\beta > 0$ and probability density function

$$f(r \mid \alpha, \beta) = \frac{\beta^\alpha}{\Gamma(\alpha)} r^{\alpha-1} e^{-\beta r},$$

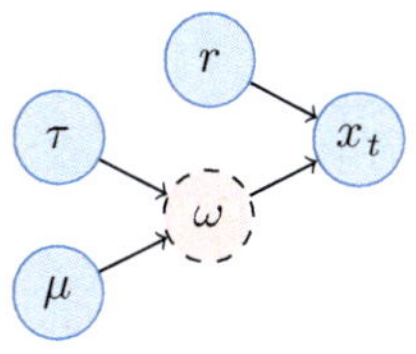

Fig. 4.7 Normal with unknown mean, graphical model

where Γ is the Gamma function. We write $r \sim \mathcal{G}amma(\alpha, \beta)$ for a random variable r that is distributed according to a Gamma distribution.

The graphical model of a Gamma distribution is shown in Fig. 4.8. The parameters α, β determine the shape and scale of the distribution, respectively. This is illustrated in Fig. 4.9, which also depicts some special cases that show that the Gamma distribution is a generalization of a number of other standard distributions. For example, for $\alpha = 1$, $\beta > 0$ one obtains an *exponential distribution* with parameter β. Its probability density function is

exponential distribution

$$f(x \mid \beta) = \beta e^{-\beta x},$$

and as the Gamma distribution it has support in $[0, \infty]$, i.e., $x > 0$. For $n \in \mathbb{N}$ and $\alpha = \frac{n}{2}$, $\beta = \frac{1}{2}$ one obtains a χ^2-distribution with n degrees of freedom.

Now we return to our problem of estimating the precision of a normal distribution with known mean, using the Gamma distribution to represent uncertainty about the precision (Fig. 4.10).

Normal-Gamma model
Let r be drawn from a Gamma distribution with parameters α, β, while x^n is a sample drawn independently from a normal distribution with mean ω and precision r, i.e.,

$$x^n \mid r \sim \mathcal{N}^n(\omega, r^{-1}), \qquad r \mid \alpha, \beta \sim \mathcal{G}amma(\alpha, \beta).$$

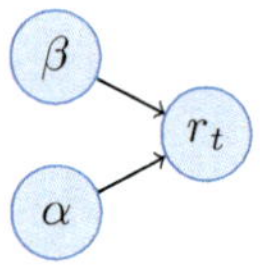

Fig. 4.8 Gamma graphical model

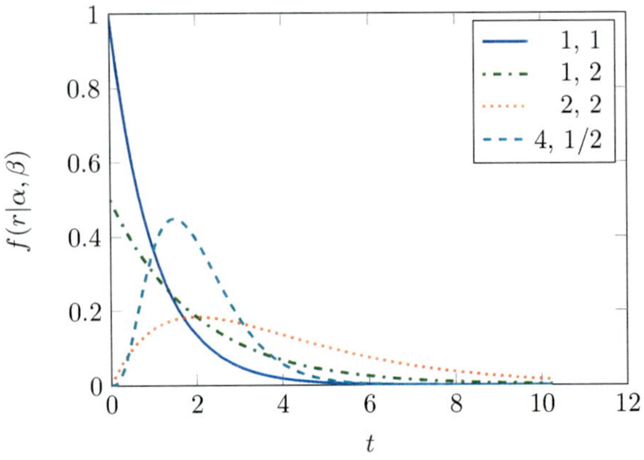

Fig. 4.9 Example Gamma densities

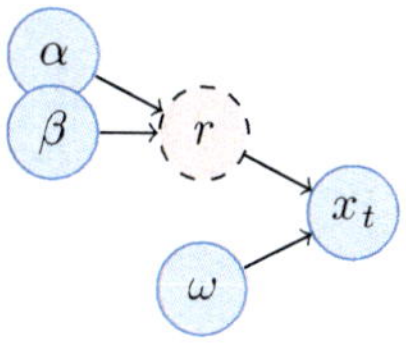

Fig. 4.10 Normal-Gamma graphical model for normal distributions with unknown precision

Then the posterior distribution of r given the sample is also Gamma, that is,

$$r \mid x^n \sim \mathcal{G}amma(\alpha', \beta') \qquad \text{with } \alpha' = \alpha + \frac{n}{2}, \qquad \beta' = \beta + \frac{1}{2}\sum_{t=1}^{n}(x_t - \omega)^2.$$

4.3.2.3 Normal with Unknown Precision and Unknown Mean

Finally, let us turn our attention to the general problem of estimating both the mean and the precision of a normal distribution. We will use the same prior distributions for the mean and precision as in the case when just one of them was unknown. It will be assumed that the precision is independent of the mean, while the mean has a normal distribution given the precision (Fig. 4.11).

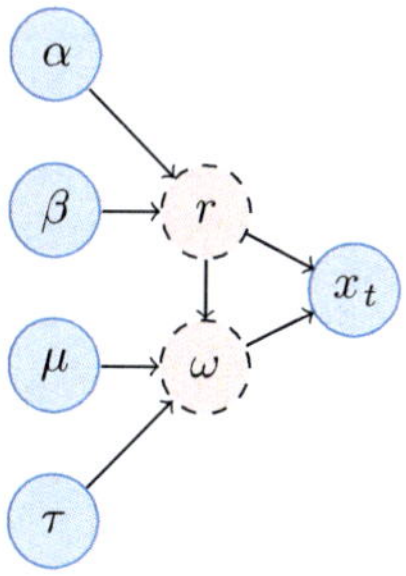

Fig. 4.11 Graphical model for a normal distribution with unknown mean and precision

Normal with unknown mean and precision
Let x^n be a sample from a normal distribution with unknown mean ω and precision r, whose prior joint distribution satisfies

$$\omega \mid r, \tau, \mu \sim \mathcal{N}(\mu, (\tau r)^{-1}), \qquad r \mid \alpha, \beta \sim \mathcal{G}amma(\alpha, \beta).$$

Then the posterior distribution is

$$\omega \mid x^n, r, \tau, \mu \sim \mathcal{N}(\mu', (\tau' r)^{-1}) \quad \text{with } r \mid x^n, \alpha, \beta \sim \mathcal{G}amma(\alpha', \beta'),$$

where

$$\mu' = \frac{\tau\mu + n\bar{x}}{\tau + n}, \qquad \tau' = \tau + n,$$

$$\alpha' = \alpha + \frac{n}{2}, \qquad \beta' = \beta + \frac{1}{2}\sum_{t=1}^{n}(x_t - \bar{x}_n)^2 + \frac{\tau n(\bar{x} - \mu)^2}{2(\tau + n)}.$$

While $\omega \mid r$ is normally distributed, the marginal distribution of ω is not normal. In fact, it can be shown that it is a *student t-distribution*. In the following, we describe the marginal distribution of a sequence of observations x^n, which is a generalized student t-distribution.

student t-distribution

The marginal distribution of x. For a normal distribution with mean ω and precision r, we have

$$f(x \mid \omega, r) \propto \sqrt{r} \cdot \exp\left(-\frac{r}{2}(x - \omega)^2\right).$$

For a prior $\omega | r \sim \mathcal{N}(\mu, (\tau r)^{-1})$ and $r \sim \mathcal{G}amma(\alpha, \beta)$, as before, the joint distribution for mean and precision is given by

$$\xi(\omega, r) \propto \sqrt{r} \cdot \exp\left(-\frac{\tau r}{2}(\omega - \mu)^2\right) r^{\alpha-1} e^{-\beta r},$$

as $\xi(\omega, r) = \xi(\omega \mid r)\,\xi(r)$. Now we can write the marginal density of new observations as

$$
\begin{aligned}
p_\xi(x) &= \int f(x \mid \omega, r)\,\mathrm{d}\xi(\omega, r) \\
&\propto \int_0^\infty \int_{-\infty}^\infty \sqrt{r}\cdot \exp\left(-\frac{r}{2}(x-\omega)^2\right) \exp\left(-\frac{\tau r}{2}(\omega-\mu)^2\right) r^{\alpha-1} e^{-\beta r}\,\mathrm{d}\omega\,\mathrm{d}r \\
&= \int_0^\infty r^{\alpha-\frac{1}{2}} e^{-\beta r} \int_{-\infty}^\infty \exp\left(-\frac{r}{2}(x-\omega)^2 - \frac{\tau r}{2}(\omega-\mu)^2\right)\mathrm{d}\omega\,\mathrm{d}r \\
&= \int_0^\infty r^{\alpha-\frac{1}{2}} e^{-\beta r} \left(\int_{-\infty}^\infty \exp\left(-\frac{r}{2}\left[(x-\omega)^2 + \tau(\omega-\mu)^2\right]\right)\mathrm{d}\omega\right)\mathrm{d}r \\
&= \int_0^\infty r^{\alpha-\frac{1}{2}} e^{-\beta r} \exp\left(-\frac{\tau r}{2(\tau+1)}(\mu-x)^2\right)\sqrt{\frac{2\pi}{r(1+\tau)}}\,\mathrm{d}r.
\end{aligned}
$$

4.3.3 Conjugates for Multivariate Distributions

The binomial distribution as well as the normal distribution can be extended to multiple dimensions. Fortunately, multivariate extensions exist for their corresponding conjugate priors, too.

4.3.3.1 Multinomial-Dirichlet Conjugates

The multinomial distribution is the extension of the binomial distribution to an arbitrary number of outcomes. It is a common model for independent random trials with a finite number of possible outcomes, such as repeated dice throws, multi-class classification problems, etc.

As in the binomial distribution we perform independent trials, but now consider a more general outcome set $\mathcal{S} = \{1, \ldots, K\}$ for each trial. Denoting by n_k the number of times one obtains outcome k, the *multinomial* distribution gives the probability of observing a given vector $(n_1, \ldots, n_K)$ after a total of n trials.

Definition 4.3.6 (*Multinomial distribution*) The multinomial distribution is discrete with parameters n, $K \in \mathbb{N}$ and a vector parameter $\boldsymbol{\omega} \in \Delta^K$, i.e., $\omega_k \geq 0$ with $\|\boldsymbol{\omega}\|_1 = 1$, where each ω_k represents the probability of obtaining outcome k in a single trial. The set of possible outcome counts after n trials is the set of vectors $\boldsymbol{n} = (n_k)_{k=1}^K$ such that $\sum_k^K n_k = n$, and the probability function is given by

$$
\mathbb{P}(\boldsymbol{n} \mid \boldsymbol{\omega}) = \frac{n!}{\prod_{k=1}^K n_k!} \prod_{k=1}^K \omega_k^{n_k}.
$$

The dependencies between the variables are shown in Fig. 4.12.

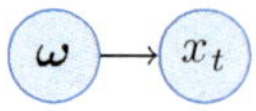

Fig. 4.12 Multinomial graphical model

4.3.3.2 The Dirichlet Distribution

The Dirichlet distribution is the multivariate extension of the Beta distribution that will turn out to be a natural candidate for a prior on the multinomial distribution.

Definition 4.3.7 (*Dirichlet distribution*) The Dirichlet distribution is a continuous distribution with outcomes $\omega \in \Omega = \mathbb{A}^K$ and a parameter vector $\boldsymbol{\alpha} \in \mathbb{R}_+^K$. Its probability density function is

$$f(\omega \mid \boldsymbol{\alpha}) = \frac{\Gamma(\sum_{i=1}^K \alpha_i)}{\prod_{k=1}^K \Gamma(\alpha_k)} \prod \omega_k^{\alpha_k - 1}.$$

We write $\omega \sim \mathcal{D}ir(\boldsymbol{\alpha})$ for a random variable ω distributed according to a Dirichlet distribution.

The parameter $\boldsymbol{\alpha}$ determines the density of the observations, as shown in Fig. 4.13.

The Dirichlet distribution is conjugate to the multinomial distribution in the same way that the Beta distribution is conjugate to the Bernoulli/binomial distribution.

Multinomial distribution with unknown parameter
Assume that the parameter ω of a multinomial distribution is generated from a Dirichlet distribution as illustrated in Fig. 4.14. If we observe $x^n = (x_1, \ldots, x_n)$, and our prior is given by $\mathcal{D}ir(\boldsymbol{\alpha})$, so that our initial belief is $\xi_0(\omega) \triangleq f(\omega \mid \boldsymbol{\alpha})$, the resulting posterior after n observations is

$$\xi_t(\omega) \propto \prod_{k=1}^K \omega_k^{n_k + \alpha_k - 1},$$

where $n_k = \sum_{t=1}^n \mathbb{I}x_t = k$.

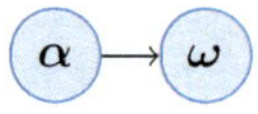

Fig. 4.13 Dirichlet graphical model

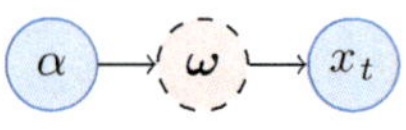

Fig. 4.14 Dirichlet-multinomial graphical model

4.3.3.3 Multivariate Normal Conjugate Families

The last conjugate pair we shall discuss is that for multivariate normal distributions. Similarly to the extension of the Bernoulli distribution to the multinomial, and the corresponding extension of the Beta to the Dirichlet, the normal priors can be extended to the multivariate case. The prior of the mean becomes a multivariate normal distribution, while that of the precision becomes a Wishart distribution.

Definition 4.3.8 (*Multivariate normal distribution*) The multivariate normal distribution is a continuous distribution with outcome space $\mathcal{S} = \mathbb{R}^K$. Its parameters are a mean vector $\boldsymbol{\omega} \in \mathbb{R}^K$ and precision matrix[3] $\boldsymbol{R} \in \mathbb{R}^{K \times K}$ that is a positive-definite, that is, $\boldsymbol{x}^\top \boldsymbol{R} \boldsymbol{x} > 0$ for $\boldsymbol{x} \neq 0$. The probability density function of the multivariate normal distribution is given by

$$f(\boldsymbol{x} \mid \boldsymbol{\omega}, \boldsymbol{R}) = (2\pi)^{-\frac{K}{2}} \sqrt{|\boldsymbol{R}|} \cdot \exp\left(-\frac{1}{2}(\boldsymbol{x}_t - \boldsymbol{\omega})^\top \boldsymbol{R}(\boldsymbol{x} - \boldsymbol{\omega})\right),$$

where $|\boldsymbol{R}|$ denotes the *matrix determinant*. When taking n independent samples from a fixed multivariate normal distribution the probability of observing a sequence $\boldsymbol{x}^n$ is given by *matrix determinant*

$$f(\boldsymbol{x}^n \mid \boldsymbol{\omega}, \boldsymbol{R}) = \prod_{t=1}^{n} f(\boldsymbol{x}_t \mid \boldsymbol{\omega}, \boldsymbol{R}).$$

The graphical model of the multivariate normal distribution is given in Fig. 4.15.

For the definition of the Wishart distribution we first have to recall the definition of a matrix trace.

Definition 4.3.9 The trace of an $n \times n$ square matrix A with entries a_{ij} is defined as

$$\mathrm{trace}(A) \triangleq \sum_{i=1}^{n} a_{ii}.$$

Definition 4.3.10 (*Wishart distribution*) The Wishart distribution is a *matrix distribution* on $\mathbb{R}^{K \times K}$ with $n > K - 1$ degrees of freedom and precision matrix $\boldsymbol{T} \in \mathbb{R}^{K \times K}$. Its probability density function is given by

$$f(\boldsymbol{V} \mid n, \boldsymbol{T}) \propto |\boldsymbol{T}|^{\frac{n}{2}} |\boldsymbol{V}|^{\frac{n-K-1}{2}} e^{-\frac{1}{2}\,\mathrm{trace}(\boldsymbol{T}\boldsymbol{V})}$$

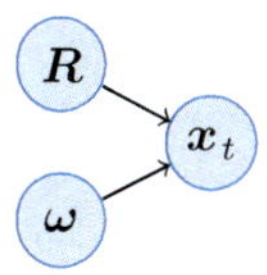

Fig. 4.15 Multivariate normal graphical model

[3] As before, the precision is the inverse of the covariance.

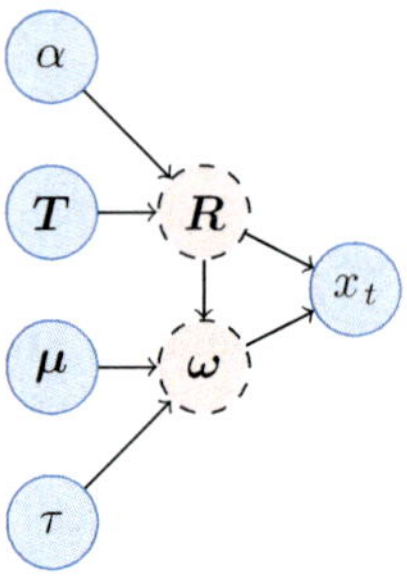

Fig. 4.16 Normal-Wishart graphical model

for positive-definite $\boldsymbol{V} \in \mathbb{R}^{K \times K}$.

Construction of the Wishart distribution. Let $\boldsymbol{x}^n$ be a sequence of n samples drawn independently from a multivariate normal distribution with mean $\boldsymbol{\omega} \in \mathbb{R}^K$ and precision matrix $\boldsymbol{T} \in \mathbb{R}^{K \times K}$, that is, $\boldsymbol{x}^n \sim \mathcal{N}^n(\boldsymbol{\omega}, \boldsymbol{T}^{-1})$. Let $\bar{\boldsymbol{x}}_n$ be the respective empirical mean, and define the covariance matrix $\boldsymbol{S} = \sum_{t=1}^n (\boldsymbol{x}_t - \bar{\boldsymbol{x}}_n)(\boldsymbol{x}_t - \bar{\boldsymbol{x}}_n)^\top$. Then $\boldsymbol{S}$ has a Wishart distribution with $n - 1$ degrees of freedom and precision matrix $\boldsymbol{T}$, and we write $\boldsymbol{S} \sim \mathcal{Wish}(n-1, \boldsymbol{T})$.

Normal-Wishart conjugate prior (Fig. 4.16)

Theorem 4.3.1 *Let $\boldsymbol{x}^n$ be a sample from a multivariate normal distribution on $\mathbb{R}^K$ with unknown mean $\boldsymbol{\omega} \in \mathbb{R}^K$ and precision $\boldsymbol{R} \in \mathbb{R}^{K \times K}$ whose joint* prior *distribution satisfies*

$$\omega \mid \boldsymbol{R} \sim \mathcal{N}(\boldsymbol{\mu}, (\tau \boldsymbol{R})^{-1}), \qquad \boldsymbol{R} \sim \mathcal{Wish}(\alpha, \boldsymbol{T}),$$

with $\tau > 0$, $\alpha > K - 1$, $\boldsymbol{T} > 0$. Then the posterior *distribution is*

$$\omega \mid \boldsymbol{R} \sim \mathcal{N}\left(\frac{\tau\boldsymbol{\mu} + n\bar{\boldsymbol{x}}_n}{\tau + n}, [(\tau + n)\boldsymbol{R}]^{-1}\right),$$

$$\boldsymbol{R} \sim \mathcal{Wish}\left(\alpha + n, \boldsymbol{T} + \boldsymbol{S} + \frac{\tau n}{\tau + n}(\boldsymbol{\mu} - \bar{\boldsymbol{x}})(\boldsymbol{\mu} - \bar{\boldsymbol{x}})^\top\right),$$

where $\boldsymbol{S} = \sum_{t=1}^n (\boldsymbol{x}_t - \bar{\boldsymbol{x}})(\boldsymbol{x}_t - \bar{\boldsymbol{x}})^\top$.

4.4 Credible Intervals

In general, according to our current belief ξ there is a certain subjective probability that some unknown parameter ω takes a certain value. However, we are not always interested in the precise probability distribution itself. Instead, we can use the complete distribution to describe an interval that we think contains the true value of the unknown parameter. In Bayesian parlance, this is called a credible interval.

Definition 4.4.1 (*Credible interval*) Given some probability measure ξ on Ω representing our belief and some interval (or set) $A \subset \Omega$,

$$\xi(A) = \int_A \mathrm{d}\xi = \mathbb{P}(\omega \in A \mid \xi)$$

is our subjective belief that the unknown parameter ω is in A. If $\xi(A) = s$, then we say that A is an s-credible interval (or set), or an interval of size (or measure) s.

As an example, for prior distributions on $\mathbb{R}$ one can construct an s-credible interval by finding $\omega_l, \omega_u \in \mathbb{R}$ such that

$$\xi([\omega_l, \omega_u]) = s.$$

Note that ω_l, ω_u are not unique and *any* choice satisfying the condition is valid. However, typically the interval is chosen so as to exclude the tails (extremes) of the distribution and centered in the maximum. Figure 4.17 shows the 90% credible interval for the Bernoulli parameter of Example 4.1 after 1000 observations, that is, the measure of A under ξ is $\xi(A) = 0.9$. We see that the true parameter $\omega = 0.6$ lies slightly outside it.

Reliability of Credible Intervals

Let ϕ, ξ_0 be probability measures on the parameter set Ω with ξ_0 being our prior belief and ϕ the actual distribution of $\omega \in \Omega$. Each ω defines a measure P_ω on the

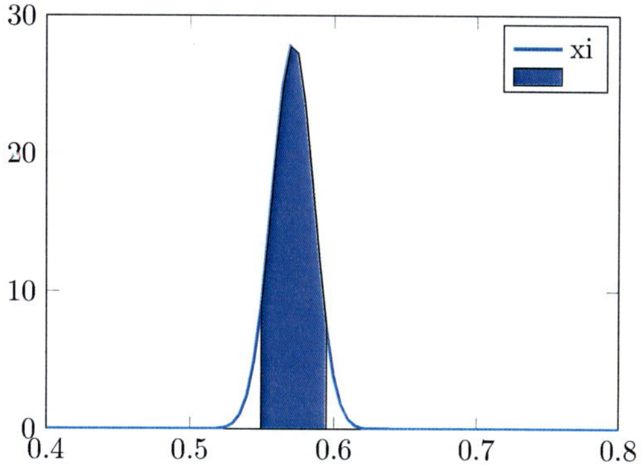

Fig. 4.17 90% credible interval after 1000 observations from a Bernoulli with $\omega = 0.6$

observation set $\mathcal{S}$. We would like to construct for any $n \in \mathbb{N}$ a credible interval $A_n \subset \Omega$ that has measure $s = \xi_n(A_n)$ after n observations. If we denote by $Q \triangleq \int_\omega P_\omega \, d\phi(\omega)$ the marginal distribution on $\mathcal{S}$, then the probability that the credible interval A_n does not include ω is

$$Q\left(\left\{x^n \in \mathcal{S}^n \mid \omega \notin A_n\right\}\right).$$

The probability that the true value of ω will be within a particular credible interval depends on how well the prior ξ_0 matches the true distribution from which the parameter ω was drawn. This is illustrated in the following experimental setup, where we check how often a 50% credible interval fails.

Experimental testing of a credible interval

Given a probability family $\mathscr{P} = \{P_\omega \mid w \in \Omega\}$.
Nature chooses distribution ϕ over Ω.
Choose distribution ξ_0 over Ω.
for $i = 1, \ldots, N$ **do**
  Draw $\omega_i \sim \phi$.
  Draw $x^n \mid \omega_i \sim P_{\omega_i}$.
  for $t = 1, \ldots, n$ **do**
    Calculate $\xi_t(\cdot) = \xi_0(\cdot \mid x^t)$.
    Construct A_t such that $\xi_t(A_t) = 0.5$.
    Check failure and set $\epsilon_{t,i} = \mathbb{I}\omega_i \notin A_t$.
  end for
end for
Average over all i, i.e., set $\epsilon_t = \frac{1}{N}\sum_{i=1}^{N} \epsilon_{t,i}$ for each t.

We performed this experiment constructing credible intervals for a Bernoulli parameter ω using $N = 1000$ trials and $n = 100$ observations per trial. Figure 4.18 illustrates what happens when we choose $\xi_0 = \phi$. We see that the credible interval is always centered around our initial mean guess and is quite tight. Figure 4.19 shows the failure rate when the credible interval A_t around our estimated mean does not match the actual value of ω_i. Since the measure of our interval A_t is always $\xi_t(A_t) = \frac{1}{2}$, we expect our error probability to be $\frac{1}{2}$, and this is borne out by the experimental results.

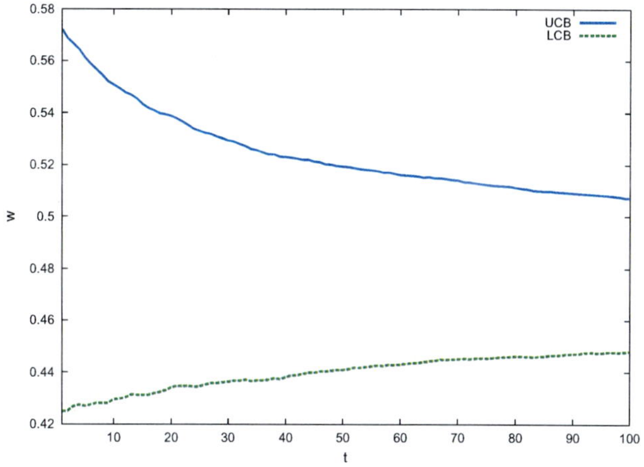

Fig. 4.18 50% credible intervals for a prior $\mathcal{B}eta(10, 10)$, ξ_0 matching the distribution of ω

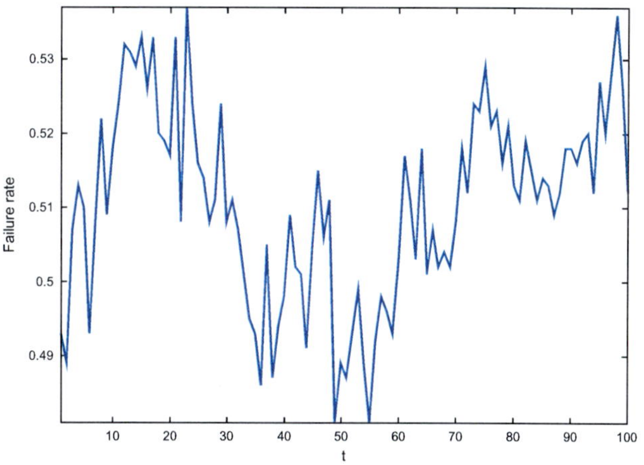

Fig. 4.19 Failure rate of 50% credible intervals for a prior $\mathcal{B}eta(10, 10)$, ξ_0 matching the distribution of ω

On the other hand, Fig. 4.20 illustrates what happens when $\xi_0 \neq \phi$. In fact, we used $\omega_i = 0.6$ for all trials i. Formally, this is a Dirac distribution concentrated in $\omega = 0.6$, usually notated as $\phi(\omega) = \delta(\omega - 0.6)$. We see that the credible interval is always centered around our initial mean guess and that it is always quite tight. Figure 4.21 shows the average number of failures. We see that initially, due to the fact that our prior is different from the true distribution, we make more mistakes than in the previous case. However, eventually, our prior is swamped by the data and the error rate converges to 50%.

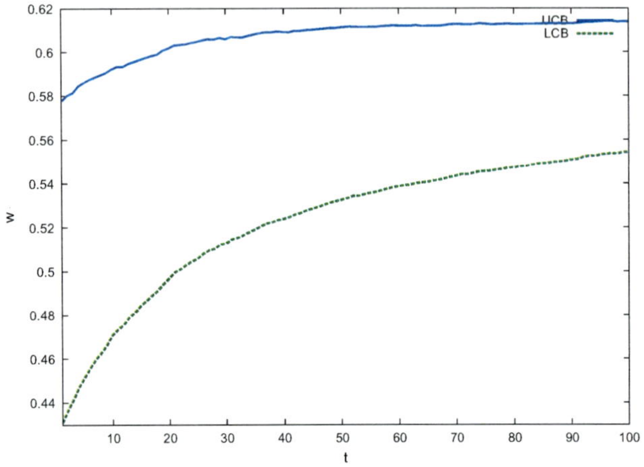

Fig. 4.20 50% credible intervals for a prior $\mathcal{B}eta(10, 10)$, when ξ_0 does not match the distribution of $\omega = 0.6$

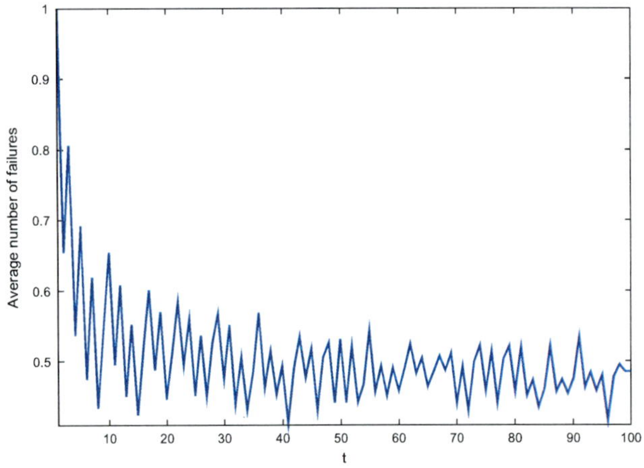

Fig. 4.21 Failure rate of 50% credible interval for a prior $\mathcal{B}eta(10, 10)$, when ξ_0 does not match the distribution of $\omega = 0.6$

4.5 Concentration Inequalities

While Bayesian ideas are useful, as they allow us to express our subjective beliefs about a particular unknown quantity, they nevertheless are difficult to employ when we have no good intuition about what prior to use. One could look at the Bayesian estimation problem as a minimax game between us and nature and consider bounds with respect to the worst possible prior distribution. However, even in that case we must select a family of distributions and priors.

In this section we will examine guarantees we can give about any calculation we make from observations with minimal assumptions about the distribution generating these observations. These findings are fundamental, in the sense that they rely on a very general phenomenon, called *concentration of measure*. As a consequence, they are much stronger than results such as the central limit theorem (which we will not cover in this textbook).

Here we shall focus on the most common application of calculating the sample mean, as given in Definition 4.2.1. We have seen that e.g. for the Beta-Bernoulli conjugate prior, it is a simple enough matter to compute a posterior distribution. From that, we can obtain a credible interval on the expected value of the unknown Bernoulli distribution. However, we would like to do the same for arbitrary distributions on $[0, 1]$, rather than just the Bernoulli. We shall now give an overview of a set of tools that can be used to do this.

Theorem 4.5.1 (Markov's inequality) *If $X \sim P$ with P a distribution on $[0, \infty)$, then*

$$\mathbb{P}(X \geq u) \leq \frac{\mathbb{E}X}{u}, \tag{4.5.1}$$

where $\mathbb{P}(X \geq u) = P\big(\{x \mid x \geq u\}\big)$.

Proof By definition of the expectation, for any u

$$\begin{aligned}
\mathbb{E}X &= \int_0^\infty x \,\mathrm{d}P(x) \\
&= \int_0^u x \,\mathrm{d}P(x) + \int_u^\infty x \,\mathrm{d}P(x) \\
&\geq 0 + \int_u^\infty u \,\mathrm{d}P(x) \\
&= u\, \mathbb{P}(X \geq u).
\end{aligned}$$

□

Consequently, if $\bar{x}_n$ is the empirical mean after n observations, for a random variable X with expectation $\mathbb{E}X = \mu$, we can use Markov's inequality to obtain $\mathbb{P}(|\bar{x}_n - \mu| \geq \epsilon) \leq \mathbb{E}|\bar{x}_n - \mu|/\epsilon$. In particular, for $X \in [0, 1]$, we obtain the bound

$$\mathbb{P}\big(|\bar{x}_n - \mu| \geq \epsilon\big) \leq \frac{1}{\epsilon}.$$

Unfortunately, this bound does not improve for a larger number of observations n. However, we can get significantly better bounds through various transformations, using Markov's inequality as a building block in other inequalities. The first of those is Chebyshev's inequality.

Theorem 4.5.2 (Chebyshev's inequality) *Let X be a random variable with expectation $\mathbb{E}X = \mu$ and variance $\mathbb{V}X = \sigma^2$. Then, for all $k > 0$,*

$$\mathbb{P}\big(|X - \mu| \geq k\sigma\big) \leq \frac{1}{k^2}. \tag{4.5.2}$$

Proof First note that for monotonic f,

$$\mathbb{P}(X \geq u) = \mathbb{P}\big(f(X) \geq f(u)\big), \tag{4.5.3}$$

as $\{x \mid x \geq u\} = \{x \mid f(x) \geq f(u)\}$. Then we can derive

$$\begin{aligned}\mathbb{P}\big(|X - \mu| \geq k\sigma\big) &= \mathbb{P}\left(\frac{|X - \mu|}{k\sigma} \geq 1\right) \overset{4.5.3}{=} \mathbb{P}\left(\frac{|X - \mu|^2}{k^2\sigma^2} \geq 1\right) \\ &\overset{4.5.1}{\leq} \mathbb{E}\left(\frac{(X - \mu)^2}{k^2\sigma^2}\right) = \frac{\mathbb{E}(X - \mu)^2}{k^2\sigma^2} = \frac{1}{k^2}.\end{aligned}$$ □

Chebyshev's inequality can in turn be used to obtain confidence bounds that improve with the number of samples n.

Example 4.3 (*Application to sample mean*) It is easy to show that the sample mean $\bar{x}_n$ has expectation μ and variance σ^2/n and we obtain from (4.5.2)

$$\mathbb{P}\left(|\bar{x}_n - \mu| \geq \frac{k\sigma}{\sqrt{n}}\right) \leq \frac{1}{k^2}.$$

Setting $\epsilon = k\sigma/\sqrt{n}$ we get $k = \epsilon\sqrt{n}/\sigma$ and hence

$$\mathbb{P}\big(|\bar{x}_n - \mu| \geq \epsilon\big) \leq \frac{\sigma^2}{\epsilon^2 n}.$$

4.5.1 Chernoff-Hoeffding Bounds

The inequality derived in Example 4.3 can be quite loose. In fact, one can prove tighter bounds for the estimation of an expected value by a different application of the Markov inequality, due to Chernoff.

Main idea of Chernoff bounds.
Let $S_n = \sum_{t=1}^{n} x_i$, with $x_t \sim P$ independently, i.e., $x^n \sim P^n$. By definition, from Markov's inequality we obtain for any $\theta > 0$

$$\begin{aligned}\mathbb{P}(S_n \geq u) &= \mathbb{P}(e^{\theta S_n} \geq e^{\theta u}) \\ &\leq e^{-\theta u}\mathbb{E}e^{\theta S_n} = e^{-\theta u}\prod_{t=1}^{n}\mathbb{E}e^{\theta x_t}.\end{aligned} \tag{4.5.4}$$

Theorem 4.5.3 (Hoeffding's inequality, [1]) *For $t = 1, 2, \ldots, n$ let $x_t \sim P_t$ be independent random variables with $x_t \in [a_t, b_t]$ and $\mathbb{E}x_t = \mu_t$. Then*

$$\mathbb{P}\left(\bar{x}_n - \mu \geq \epsilon\right) \leq \exp\left(-\frac{2n^2\epsilon^2}{\sum_{t=1}^{n}(b_t - a_t)^2}\right), \tag{4.5.5}$$

where $\mu = \frac{1}{n}\sum_{t=1}^{n}\mu_t$.

Proof Applying (4.5.4) to random variables $x_t - \mu_t$ so that $S_n = n(\bar{x}_n - \mu)$ and setting $u = n\epsilon$ gives

$$\begin{aligned}\mathbb{P}(\bar{x}_n - \mu \geq \epsilon) &= \mathbb{P}(S_n \geq u) \\ &\leq e^{-\theta n\epsilon}\prod_{t=1}^{n}\mathbb{E}e^{\theta(x_t - \mu_t)}.\end{aligned} \tag{4.5.6}$$

Applying Jensen's inequality directly to the expectation does not help. However, we can use convexity in another way: Let $f(z)$ be the following linear upper bound on $e^{\theta z}$ on the interval $[a, b]$:

$$f(z) \triangleq \frac{b - z}{b - a}e^{\theta a} + \frac{z - a}{b - a}e^{\theta b} \geq e^{\theta z}$$

Then obviously $\mathbb{E}e^{\theta z} \leq \mathbb{E}f(z)$ for $z \in [a, b]$. We can use this to bound the term inside the product in (4.5.6), setting $z = x_t - \mu_t$:

$$e^{\theta(x_t - \mu_t)} \leq \frac{e^{\ \theta\mu_t}}{b_t - a_t}\left[(b_t - \mu_t)e^{\theta a_t} + (\mu_t - a_t)e^{\theta b_t}\right]$$

Bounding the expectation of this term by taking derivatives with respect to θ and computing the second order Taylor expansion gives

$$\mathbb{E}e^{\theta(x_t - \mu_t)} \leq e^{\frac{1}{8}\theta^2(b_t - a_t)^2}.$$

We conclude by plugging in this bound in (4.5.6):

$$\mathbb{P}(\bar{x}_n - \mu \geq \epsilon) \leq e^{-\theta n\epsilon + \frac{1}{8}\theta^2 \sum_{t=1}^n (b_t - a_t)^2}.$$

This is minimized for $\theta = 4n\epsilon / \sum_{t=1}^n (b_t - a_t)^2$, which proves the required result. □

We can apply this inequality directly to the sample mean example and obtain for $x_t \in [0, 1]$

$$\mathbb{P}(|\bar{x}_n - \mu| \geq \epsilon) \leq 2e^{-2n\epsilon^2}.$$

4.6 Approximate Bayesian Approaches

Unfortunately, exact computation of the posterior distributions is only possible in special cases. In this section, we give a brief overview of some classic methods for approximate Bayesian inference. The first, Monte Carlo methods, rely on stochastic approximations of the posterior distributions where at least the likelihood function is computable. The second, approximate Bayesian computation, extends Monte Carlo methods to the case where the probability function is incomputable or not available at all. In the third, which includes variational Bayes methods, we replace distributions with an analytic approximation. Finally, in empirical Bayes methods, some parameters are replaced by an empirical estimate.

4.6.1 Monte Carlo Inference

Monte Carlo inference has been a cornerstone of approximate Bayesian statistics ever since computing power was sufficient for such methods to become practical. Let us begin with a simple example, that of estimating expectations.

Definition 4.6.1 Let $f : \mathcal{S} \to [0, 1]$ and P be a measure on $\mathcal{S}$. Then

$$\mathbb{E}_P f \triangleq \int_{\mathcal{S}} f(x) \, \mathrm{d}P(x).$$

Estimating the expectation $\mathbb{E}_P f$ is relatively easy as long as we can generate samples from P. Then a simple average provides an estimate that can be computed fast and whose error can be bounded by Hoeffding's inequality.

Corollary 4.6.1 *Let $x^n = (x_1, \ldots, x_n)$ be a sample of size n with $x_t \sim P$ and $f : \mathcal{S} \to [0, 1]$. Setting $\hat{f}_n \triangleq \frac{1}{n}\sum_t f(x_t)$ it holds that*

$$P\left(\left\{x^n \in \mathcal{S}^n \mid |\hat{f}_n - \mathbb{E} f| \geq \epsilon\right\}\right) \leq 2e^{-2n\epsilon^2}.$$

This technique can also be used to calculate posterior distributions.

Example 4.4 (*Calculation of posterior distributions*) Assume a probability family $\mathscr{P} = \{P_\omega \mid \omega \in \Omega\}$ and a prior distribution ξ on Ω such that we can draw $\omega \sim \xi$. The posterior distribution is given by (4.1.1), where we can write the nominator as

$$\int_B P_\omega(x)\, \mathrm{d}\xi(\omega) = \int_\Omega \mathbb{I}\{\omega \in B\} P_\omega(x)\, \mathrm{d}\xi(\omega) = \mathbb{E}_\xi \left[\mathbb{I}\{\omega \in B\} P_\omega(x)\right]. \quad (4.6.1)$$

Similarly, the denominator of (4.1.1) can be written as $\mathbb{E}_\xi[P_\omega(x)]$. If P_ω is bounded, then the error can be bounded too.

An extension of this approach involves Markov chain Monte Carlo (MCMC) methods. These are sequential sampling procedures where data are sampled iteratively. At the t-th iteration, we obtain a sample $x_t \sim Q_t$, where Q_t depends on the previous sample x_{t-1}. Even if one can guarantee that $Q_t \to P$, there is no easy way to determine *a priori* when the procedure has converged. For more details see for example [2].

4.6.2 Approximate Bayesian Computation

The main problem approximate Bayesian computation (ABC) aims to solve is how to weight the evidence we have for or against different models. The assumption is that we have a family of models $\{M_\omega \mid \omega \in \Omega\}$, from which we can generate data. However, there is no easy way to calculate the probability of any model having generated the data. On the other hand, like in the standard Bayesian setting, we can start with a prior ξ over Ω and given some data $x \in \mathcal{W}$ we wish to calculate the posterior $\xi(\omega \mid x)$. ABC methods generally rely on what is called an *approximate statistic* in order to weight the relative likelihood of models given the data.

An approximate statistic $\phi : \mathcal{S} \to \mathcal{S}'$ maps the data to some lower dimensional space $\mathcal{S}'$. Then it is possible to compare different data points in terms of how similar their statistics are. For this, one also defines some distance $D : \mathcal{S}' \times \mathcal{S}' \to \mathbb{R}_+$.

ABC methods are useful in two specific situations. The first is when the family of models that we consider has an intractable likelihood. This means that calculating $M_\omega(x)$ is prohibitively expensive. The second is in some applications which admit a class of *parametrized simulators* which have no probabilistic description. Then one reasonable approach is to find the best simulator in the class and apply it to the actual problem.

The simplest algorithm in this context is ABC Rejection Sampling shown as Algorithm 4.1. Here, we repeatedly sample a model from the prior distribution, and then generate data $\hat{x}$ from the model. If the sampled data is ϵ-close to the original data in terms of the statistic, we accept the sample as an approximate posterior sample.

For an overview of ABC methods see [3, 4]. Early ABC methods were developed for applications such as econometric modelling e.g. [5], where detailed simulators

Algorithm 4.1 ABC Rejection Sampling from $\xi(\omega \mid x)$.

input prior ξ, data x, generative model family $\{M_\omega \mid \omega \in \Omega\}$, statistic ϕ, error bound ϵ.
repeat
  $\hat{\omega} \sim \xi$
  $\hat{x} \sim M_{\hat{\omega}}$
until $D[\phi(x), \phi(\hat{x})] \leq \epsilon$
Return $\hat{\omega}$

but no useful analytical probabilistic models were available. ABC methods have also been used for inference in dynamical systems e.g. [6] and the reinforcement learning problem [7, 8].

4.6.3 Analytic Approximations of the Posterior

Another type of approximation involves substituting complex distributions with members from a simpler family. For example, one could replace a multimodal posterior distribution $\xi(\omega \mid x)$ with a Gaussian. A more principled approximation would involve selecting a distribution that is the closest with respect to some divergence or distance. In particular, we would like to approximate the target distribution $\xi(\omega \mid x)$ with some other distribution $Q_\theta(\omega)$ in a family $\{Q_\theta \mid \theta \in \Theta\}$ such that the distance between $\xi(\omega \mid x)$ and $Q_\theta(\omega)$ is minimal. For measuring the latter one can use the total variation or the Wasserstein distance. The most popular algorithms employ however the KL-divergence

$$D(Q \parallel P) \triangleq \int_\Omega \ln \frac{\mathrm{d}Q}{\mathrm{d}P} \, \mathrm{d}Q.$$

As the KL-divergence is asymmetric, its use results in two distinct approximation methods, variational Bayes and expectation propagation.

Variational approximation. In this formulation, we wish to minimize the KL-divergence

$$D\left(Q_\theta \parallel \xi_{|x}\right) = \int_\Omega \ln \frac{\mathrm{d}Q_\theta}{\mathrm{d}\xi_{|x}} \, \mathrm{d}Q_\theta,$$

where $\xi_{|x}$ is shorthand for the distribution $\xi(\omega \mid x)$. An efficient method for minimizing this divergence is rewriting it as

$$D\left(Q_\theta \parallel \xi_{|x}\right) = -\int_\Omega \ln \frac{d\xi_{|x}}{dQ_\theta}\, dQ_\theta$$
$$= -\int_\Omega \ln \frac{d\xi_{|x}}{dQ_\theta}\, dQ_\theta + \ln \xi(x),$$

where $\xi_{|x}$ is shorthand for the joint distribution $\xi(\omega, x)$ for a fixed value of x. As the second term does not depend on θ, we can find the best element of the family by computing

$$\max_{\theta\in\Theta} \int_\Omega \ln \frac{d\xi_{|x}}{dQ_\theta}\, dQ_\theta,$$

where the term we are maximizing can also be seen as a lower bound on the marginal log-likelihood.

Expectation propagation. The other direction requires us to minimize the divergence

$$D\left(\xi_{|x} \parallel Q_\theta\right) = \int_\Omega \ln \frac{d\xi_{|x}}{dQ_\theta}\, d\xi_{|x}.$$

A respective algorithm in the case of data terms that are independent given the parameter is expectation propagation [9]. There, the approximation has a factored form and is iteratively updated, with each term minimizing the KL-divergence while keeping the remaining terms fixed.

4.6.4 Maximum Likelihood and Empirical Bayes Methods

When it is not necessary to have a full posterior distribution, some parameter may be estimated point-wise. One simple such approach is maximum likelihood. In the simplest case, we replace the posterior distribution $\xi(\omega \mid x)$ with a point estimate corresponding to the parameter value that maximizes the likelihood, that is,

$$\omega^*_{\mathrm{ML}} \in arg\,\max_\omega P_\omega(x).$$

Alternatively, the *maximum a posteriori* parameter may be obtained by

$$\omega^*_{\mathrm{MAP}} \in arg\,\max_\omega \xi(\omega \mid x).$$

In the latter case, even though we cannot compute the full function $\xi(\omega \mid x)$, we can still maximize (perhaps locally) for ω.

More generally, there might be some parameters ϕ for which we actually *can* compute a posterior distribution, and some other parameters ω for which we can not. Then we can perform Bayesian inference for the ϕ parameters and maximum-likelihood for the ω parameters. This generally falls within the domain of Empirical Bayes methods, pioneered by [10]. These replace some parameters by an empirical estimate not necessarily corresponding to the maximum likelihood. These methods are quite diverse [10–14] and unfortunately beyond the scope of this book.

References

1. Hoeffding, W.: Probability inequalities for sums of bounded random variables. J. Am. Stat. Assoc. **58**(301), 13–30 (1963)
2. Casella, G., Fienberg, S., Olkin, I. (eds.) Monte Carlo Statistical Methods. Springer Texts in Statistics. Springer (1999)
3. Csilléry, K., Blum, M.G.B., Gaggiotti, O.E., François, O.: Approximate Bayesian computation (ABC) in practice. Trends Ecol. Evol. **25**(7), 410–418 (2010)
4. Marin, J.-M., Pudlo, P., Robert, C.P., Ryder, R.J.: Approximate Bayesian computational methods. Stat. Comput. **22**(6), 1167–1180 (2012)
5. Geweke, J.F.: Using simulation methods for Bayesian econometric models: Inference, development, and communication. Econ. Rev. **18**(1), 1–73 (1999)
6. Toni, T., Welch, D., Strelkowa, N., Ipsen, A., Stumpf, M.P.H.: Approximate Bayesian computation scheme for parameter inference and model selection in dynamical systems. J. Royal Soc. Interf. **6**(31), 187–202 (2009)
7. Dimitrakakis, C., Tziortziotis, N.: ABC reinforcement learning. In: Proceedings of the 30th International Conference on Machine Learning, ICML 2013, pp. 684–692. JMLR.org (2013)
8. Dimitrakakis, C., Tziortziotis, N.: Usable ABC reinforcement learning. In: NIPS 2014 Workshop: ABC in Montreal (2014)
9. Minka, T.P.: Expectation propagation for approximate Bayesian inference. In: UAI '01: Proceedings of the 17th Conference in Uncertainty in Artificial Intelligence, pp. 362–369. Morgan Kaufmann (2001)
10. Robbins, H.: An empirical Bayes approach to statistics. In: Neyman, J. (ed.) Proceedings of the Third Berkeley Symposium on Mathematical Statistics and Probability, Volume 1: Contributions to the Theory of Statistics. University of California Press, Berkeley, CA (1955)
11. Laird, N.M., Louis, T.A.: Empirical Bayes confidence intervals based on bootstrap samples. J. Am. Stat. Assoc. **82**(399), 739–750 (1987)
12. Lwin, T., Maritz, J.S.: Empirical Bayes approach to multiparameter estimation: with special reference to multinomial distribution. Ann. Inst. Stat. Math. **41**(1), 81–99 (1989)
13. Robbins, H.: The empirical Bayes approach to statistical decision problems. Ann. Math. Stat. **35**(1), 1–20 (1964)
14. Deely, J.J., Lindley, D.V.: Bayes empirical Bayes. J. Am. Stat. Assoc. **76**(376), 833–841 (1981)

Chapter 5
Sequential Sampling

5.1 Gains From Sequential Sampling

So far, we have mainly considered decision problems where the sample size was fixed. However, frequently the sample size can also be part of the decision. Since normally larger sample sizes give us more information, in this case the decision problem is only interesting when obtaining new samples has a cost. Consider the following example.

Example 5.1 Consider that you have 100 produced items and you want to determine whether there are fewer than 10 faulty items among them. If testing has some cost, it pays off to think about whether it is possible to do without testing all 100 items. Indeed, this is possible by the following simple online testing scheme: You test one item after another until you either have discovered 10 faulty items or 91 good items. In either case you have the correct answer at considerably lower cost than when testing all items.

A sequential sample from some unknown distribution P is generated as follows. First, let us fix notation and assume that each new sample x_i we obtain belongs to some alphabet $\mathcal{X}$, so that at time t, we have observed $x_1, \ldots, x_t \in \mathcal{X}^t$. It is also convenient to define the set of all sequences in the alphabet $\mathcal{X}$ as $\mathcal{X}^* \triangleq \bigcup_{t=0}^{\infty} \mathcal{X}^t$. The distribution P defines a probability on $\mathcal{X}^*$ so that x_{t+1} may depend on the previous samples $x_1, \ldots, x_t$ in an arbitrary manner. At any time t, we can either *stop sampling* or obtain one *more* observation x_{t+1}. A sample obtained in this way is called a *sequential sample*. More formally, we give the following definition:

Definition 5.1.1 (*Sequential sampling*) A sequential sampling procedure on a probability space[1] $(\mathcal{X}^*, \mathfrak{B}(\mathcal{X}^*), P)$ involves a *stopping function* $\pi_s : \mathcal{X}^* \to \{0, 1\}$, such that we stop sampling at time t if and only if $\pi_s(x^t) = 1$, otherwise we obtain a new sample $x_{t+1} \mid x^t \sim P(\cdot \mid x^t)$. *stopping function*

[1] This is simply a sample space and associated algebra, together with a probability measure.

C. Dimitrakakis and R. Ortner, *Decision Making Under Uncertainty and Reinforcement Learning*, Intelligent Systems Reference Library 223,
https://doi.org/10.1007/978-3-031-07614-5_5

Thus, the sample obtained depends both on P and the sampling procedure π_s. In our setting, we don't just want to sample sequentially, but also to take some action after sampling is complete. For that reason, we can generalize the above definition to sequential decision procedures.

Definition 5.1.2 (*Sequential decision procedure*) A sequential decision procedure $\pi = (\pi_s, \pi_d)$ is a tuple composed of

1. a stopping rule $\pi_s : \mathcal{X}^* \to \{0, 1\}$ and
2. a decision rule $\pi_d : \mathcal{X}^* \to \mathcal{A}$.

The stopping rule π_s specifies whether, at any given time, we should stop and make a decision in $\mathcal{A}$ or take one more sample. That is, stop if

$$\pi_s(x^t) = 1,$$

otherwise observe x_{t+1}. Once we have stopped (i.e. $\pi_s(x^t) = 1$), we choose the decision

$$\pi_d(x^t).$$

Deterministic stopping rules If the stopping rule π_s is deterministic, then for any t, there exists some *stopping set* $B_t \subset \mathcal{X}^t$ such that

stopping set

$$\pi_s(x^t) = \begin{cases} 1, & \text{if } x^t \in B_t \\ 0, & \text{if } x^t \notin B_t. \end{cases} \tag{5.1}$$

As with any Bayesian decision problem, it is sufficient to consider only deterministic decision rules.

We are interested in sequential sampling problems especially when there is a reason for us to stop sampling early enough, like in the case when we incur a cost with each sample we take. A detailed example is given in the following section.

5.1.1 An Example: Sampling with Costs

We once more consider problems where we have some observations $x_1, x_2, \ldots$, with $x_t \in \mathcal{X}$, which are drawn from some distribution with parameter $\theta \in \Theta$, or more precisely from a family $\mathscr{P} = \{P_\theta \mid \theta \in \Theta\}$, such that each $(\mathcal{X}, \mathfrak{B}(\mathcal{X}), P_\theta)$ is a probability space for all $\theta \in \Theta$. Since we take repeated observations, the probability of a sequence $x^n = x_1, \ldots, x_n$ under an i.i.d. model θ is $P_\theta^n(x^n)$. We have a prior probability measure ξ on $\mathfrak{B}(\Theta)$ for the unknown parameter, and we wish to take an action $a \in \mathcal{A}$ that maximizes the expected utility according to a utility function $u : \Theta \times \mathcal{A} \to \mathbb{R}$.

In the classical case, we obtain a complete sample of fixed size n, $x^n = (x_1, \ldots, x_n)$ and calculate a posterior measure $\xi(\cdot \mid x^n)$. We then take the decision maximizing the expected utility according to our posterior. Now consider the case of sampling with costs, such that a sample of size n results in a cost of cn. For that reason we define a new utility function U which depends on the number of observations we have.

Samples with costs

$$U(\theta, a, x^n) = u(\theta, a) - cn, \tag{5.2}$$

$$\mathbb{E}_\xi(U \mid a, x^n) = \int_\Theta u(\theta, a) \, d\xi(\theta \mid x^n) - cn. \tag{5.3}$$

In the remainder of this section, we shall consider the following simple decision problem, where we need to make a decision about the value of an unknown parameter. As we get more data, we have a better chance of discovering the right parameter. However, there is always a small chance of getting no information.

Example 5.2 Consider the following decision problem, where the goal is to distinguish between two possible hypotheses θ_1, θ_2, with corresponding decisions a_1, a_2. We have three possible observations $\{1, 2, 3\}$, with $1, 2$ being more likely under the first and second hypothesis, respectively. However, the third observation gives us no information about the hypothesis, as its probability is the same under θ_1 and θ_2. In this problem γ is the probability that we obtain an uninformative sample.

- Parameters: $\Theta = \{\theta_1, \theta_2\}$.
- Decisions: $\mathcal{A} = \{a_1, a_2\}$.
- Observation distribution $f_i(k) = \mathbb{P}_{\theta_i}(x_t = k)$ for all t with

$$f_1(1) = 1 - \gamma, \qquad f_1(2) = 0, \qquad f_1(3) = \gamma, \tag{5.4}$$

$$f_2(1) = 0, \qquad f_2(2) = 1 - \gamma, \qquad f_2(3) = \gamma. \tag{5.5}$$

- Local utility: $u(\theta_i, a_j) = 0$, for $i = j$ and $b < 0$ otherwise.
- Prior: $P_\xi(\theta = \theta_1) = \xi = 1 - P_\xi(\theta = \theta_2)$.
- Observation cost per sample: c.

At any step t, you have the option of continuing for one more step, or stopping and taking an action in $\mathcal{A}$. The question is what is the policy for sampling and selecting an action that maximizes expected utility?

In this problem, it is immediately possible to distinguish θ_1 from θ_2 when you observe $x_t = 1$ or $x_t = 2$. However, the values $x_t = 3$ provide no information. The expected utility of stopping if you have only observed 3's after t steps is $\xi b - ct$. In fact, if your posterior parameter after t steps is ξ_t, then the expected utility of stopping is $b \min\{\xi_t, 1 - \xi_t\} - ct$. In general, you should expect ξ_t to approach 0 or 1 with high probability, and hence taking more samples is better. However, if we pay utility $-c$ for each additional sample, there is a point of diminishing returns, after which it will not be worthwhile to take any more samples.

value We first investigate the setting where the number of observations is fixed. In particular, the *value* of the optimal procedure taking n observations is defined to be the expected utility that maximizes the *a posteriori* utility given x^n, i.e.,

$$V(n) = \sum_{x^n} P^n_\xi(x^n) \max_a \mathbb{E}_\xi(U \mid x^n, a),$$

where $P^n_\xi = \int_\Theta P^n_\theta \, d\xi(\theta)$ is the marginal distribution over n observations. For this specific example, it is easy to calculate the value of the procedure that takes n observations, by noting the following facts.

(a) The probability of observing $x_t = 3$ for all $t = 1, \ldots, n$ is γ^n. Then we must rely on our prior ξ to make a decision.
(b) If we observe any other sequence, we know the value of θ.

Consequently, the total value $V(n)$ of the optimal procedure taking n observations is

$$V(n) = \xi b \gamma^n - cn.$$

Based on this, we now want to find the optimal number of samples n. Since V is a smooth function, an approximate maximizer can be found by viewing n as a continuous variable.[2] Taking derivatives, we get

$$n^* = \left[\log \frac{c}{\xi b \log \gamma}\right] \frac{1}{\log \gamma},$$

$$V(n^*) = \frac{c}{\log \gamma} \left[1 + \log \frac{c}{\xi b \log \gamma}\right].$$

The results of applying this procedure are illustrated in Fig. 5.1. Here we can see that, for two different choices of priors, the optimal number of samples is different. In both cases, there is a clear choice for how many samples to take, when we must fix the number of samples before seeing any data.

However, we may *not* be constrained to fix the number of samples a priori. As illustrated in Example 5.1, many times it is a good idea to adaptively decide when to stop taking samples. This is illustrated by the following *sequential* procedure. Since we already know that there is an optimal *a priori* number of steps n^*, we can choose to look at all possible stopping times that are smaller or equal to n^*.

---

[2] In the end, we can find the optimal maximizer by looking at the nearest two integers to the value found.

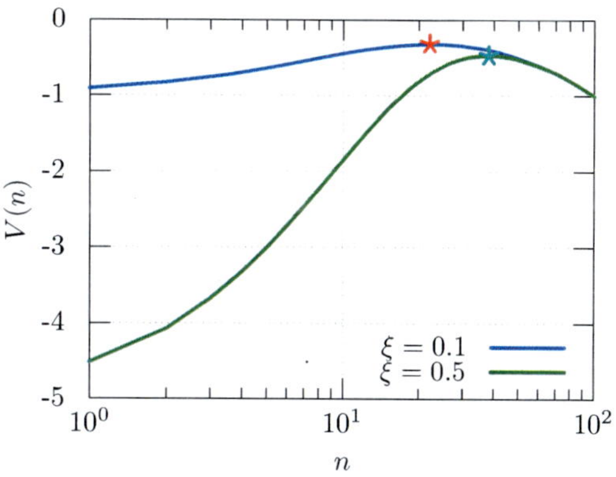

Fig. 5.1 Illustration of P1, the procedure taking a fixed number of samples n. The value of taking exactly n observations under two different beliefs, for $\gamma = 0.9$, $b = -10$, $c = 10^{-2}$

P2. A sequential procedure stopping after at most n^* steps

- If $t < n^*$, use the stopping rule $\pi_s(x^t) = 1$ iff $x_t \neq 3$.
- If $t = n^*$, then stop.
- Our posterior after stopping is $\xi(\theta \mid x^n)$, where both x^n and the number of observations n are random.

Since the probability of $x_t = 3$ is always the same for both θ_1 and θ_2, we have

$$\mathbb{E}_\xi(n) = \mathbb{E}(n \mid \theta = \theta_1) = \mathbb{E}(n \mid \theta = \theta_2) < n^*.$$

We can calculate the expected number of observations as

$$\begin{aligned}\mathbb{E}_\xi(n \mid n \leq n^*) = \mathbb{E}_\xi(n \mid \theta = \theta_1) &= \sum_{t=1}^{n^*} t\, \mathbb{P}_\xi(n = t \mid \theta = \theta_1) \\ &= \sum_{t=1}^{n^*-1} t\gamma^{t-1}(1-\gamma) + n^*\gamma^{n^*-1} = \frac{1-\gamma^{n^*}}{1-\gamma},\end{aligned}$$

using the formula for the *geometric series*. Consequently, the value of this procedure is

$$\begin{aligned}\bar{V}(n^*) &= \mathbb{E}_\xi(U \mid n = n^*)\, \mathbb{P}_\xi(n = n^*) + \mathbb{E}_\xi(U \mid n < n^*)\, \mathbb{P}_\xi(n < n^*) \\ &= \xi b \gamma^{n^*} - c\, \mathbb{E}_\xi(n),\end{aligned}$$

and from the definition of n^* we obtain

$$\bar{V}(n^*) = \frac{c}{\gamma - 1} + \frac{c}{\log \gamma}\left[1 + \frac{c}{\xi b(1-\gamma)}\right].$$

Fig. 5.2 The value of three strategies for $\xi = 1/2$, $b = -10$, $c = 10^{-2}$ and varying γ. Higher values of γ imply a longer time before the true θ is known

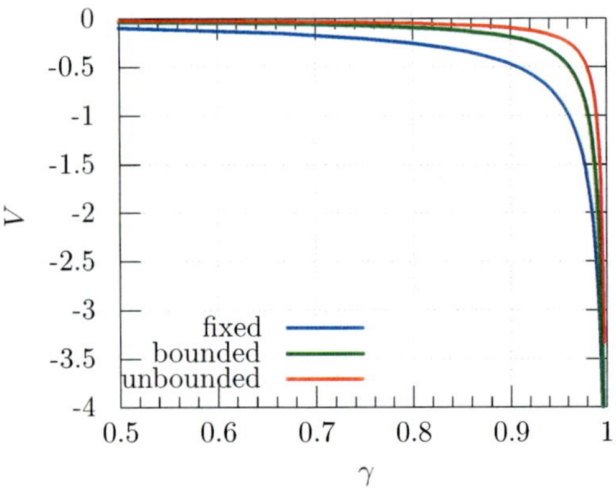

As you can see, there is a non-zero probability that $n = n^*$, at which time we will have not resolved the true value of θ. In that case, we are still not better off than at the very beginning of the procedure, when we had no observations. If our utility is linear with the number of steps, it thus makes sense that we should still continue. For that reason, we should consider *unbounded procedures*.

unbounded procedures

The unbounded procedure for our example is simply this to use the stopping rule $\pi_s(x^t) = 1$ iff $x_t \neq 3$. Since we only obtain information whenever $x_t \neq 3$, and that information is enough to fully decide θ, once we observe $x_t \neq 3$, we can make a decision that has value 0, as we can guess correctly. So, the value of the unbounded sequential procedure is just $V^* = -c\,\mathbb{E}_\xi(n)$, where

$$\mathbb{E}_\xi(n) = \sum_{t=1}^{\infty} t\,\mathbb{P}_\xi(n = t) = \sum_{t=1}^{\infty} t\gamma^{t-1}(1-\gamma) = \frac{1}{1-\gamma},$$

again using the formula for the geometric series.

In the given example, it is clear that bounded procedures are (in expectation) better than fixed-sampling procedures, as seen in Fig. 5.2. In turn, the unbounded procedure is (in expectation) better than the bounded procedure. Of course, an unbounded procedure may end up costing much more than taking a decision without observing any data, as it disregards the costs up to time t. This relates to the economic idea of *sunk costs*: since our utility is additive in terms of the cost, our optimal decision now should not depend on previously accrued costs.

5.2 Optimal Sequential Sampling Procedures

We now turn our attention to the general case. While it is easy to define the optimal stopping rule and decision in this simple example, how can we actually do the same thing for *arbitrary* problems? The following section characterizes optimal sequential sampling procedures and gives an algorithm for constructing them.

Once more, consider a distribution family $\mathscr{P} = \{P_\theta \mid \theta \in \Theta\}$ and a prior ξ over $\mathfrak{B}(\Theta)$. For a decision set $\mathcal{A}$, a utility function $U : \Theta \times \mathcal{A} \to \mathbb{R}$, and sampling costs c, the utility of a sequential decision procedure π is the local utility at the end of the procedure, minus the sampling cost. In expectation, this can be written as

$$U(\xi, \pi) = \mathbb{E}_\xi \left\{u[\theta, \pi(x^n)] - nc\right\}.$$

Here the cost is inside the expectation, since the number of samples we take is random. Summing over all the possible stopping times n, and taking $B_n \subset \mathcal{X}^*$ as the set of observations for which we stop, we have

$$\begin{aligned} U(\xi, \pi) &= \sum_{n=1}^{\infty} \int_{B_n} \mathbb{E}_\xi[U(\theta, \pi(x^n)) \mid x^n] \, \mathrm{d}P_\xi(x^n) - \sum_{n=1}^{\infty} P_\xi(B_n) nc \\ &= \sum_{n=1}^{\infty} \int_{B_n} \left\{ \int_\Theta U[\theta, \pi(x^n)] \, \mathrm{d}\xi(\theta \mid x^n) \right\} \mathrm{d}P_\xi(x^n) - \sum_{n=1}^{\infty} P_\xi(B_n) nc, \end{aligned}$$

where P_ξ is the marginal distribution under ξ. Although it may seem difficult to evaluate this, it can be done by a simple dynamic programming technique called *backwards induction*. We first give the algorithm for the case of bounded procedures (i.e. procedures that must stop after a particular time) and later for unbounded ones. *backwards induction*

Definition 5.2.1 (*Bounded sequential decision procedure*) A sequential decision procedure is T-*bounded* for a positive integer T if $\mathbb{P}_\xi(n \leq T) = 1$. The procedure is called *bounded* if it is T-bounded for some T.

We can analyse such a procedure by recursively analysing procedures for larger T, starting from the final point of the process and working our way backwards. Consider a π that is T-bounded. Then we know that we shall take at most T samples. If the process ends at stage T, we will have observed some sequence x^T, which gives rise to a posterior $\xi(\theta \mid x^T)$. Since we *must* stop at T, we must choose a maximizing expected utility at that stage, that is,

$$\mathbb{E}_\xi[U \mid x^T, a] = \int_\Theta U(\theta, a) \, \mathrm{d}\xi(\theta \mid x^T).$$

Since we need not take another sample, the respective value (maximal expected utility) of that stage is

$$V^0[\xi(\cdot|x^T)] \triangleq \max_{a \in \mathcal{A}} U(\xi(\cdot \mid x^T), a),$$

where we introduce the notation V^n to denote the expected utility, given that we are stopping after at most n steps.

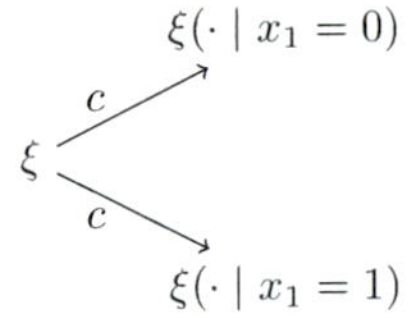

Fig. 5.3 An example of a sequential decision problem with two stages. The initial belief is ξ and there are two possible subsequent beliefs, depending on whether we observe $x_t = 0$ or $x_t = 1$. At each stage we pay c

More generally, we need to consider the effect on subsequent decisions. Consider the following simple two-stage problem as an example. Let $\mathcal{X} = \{0, 1\}$ and ξ be the prior on the θ parameter of $Bern(\theta)$. We wish to either decide immediately on a parameter θ, or take one more observation, at cost c, before deciding. The problem we consider has two stages, as illustrated in Fig. 5.3.

In this example, we begin with a prior ξ at the first stage. There are two possible outcomes for the **second stage**.

1. If we observe $x_1 = 0$ then our value is $V^0[\xi(\cdot \mid x_1 = 0)]$.
2. If we observe $x_1 = 1$ then our value is $V^0[\xi(\cdot \mid x_1 = 1)]$.

At the first stage, we can:

1. Stop with value $V^0(\xi)$.
2. Pay a sampling cost c for value $V^0[\xi(\cdot \mid x_1)]$ with $P_\xi(x_1) = \int_\Theta P_\theta(x_1) \, d\xi(w)$.

So the expected value of continuing for one more step is

$$V^1(\xi) \triangleq \int_{\mathcal{X}} V^0[\xi(\cdot \mid x_1)] \, dP_\xi(x_1).$$

Thus, the overall value for this problem is:

$$\max \left\{ V^0(\xi), \sum_{x_1=0}^{1} V^0[\xi(\cdot \mid x_1)] P_\xi(x_1) - c \right\}$$

The above is simply the maximum of the value of stopping immediately (V^0), and the value of continuing for at most one more step (V^1). This procedure can be applied recursively for multi-stage problems, as explained below.

5.2.1 Multi-stage Problems

For simplicity, we use ξ_n to denote a posterior $\xi(\cdot \mid x^n)$, omitting the specificbreak-value of x^n. For any specific ξ_n, there is a range of given possible next beliefs ξ_{n+1},

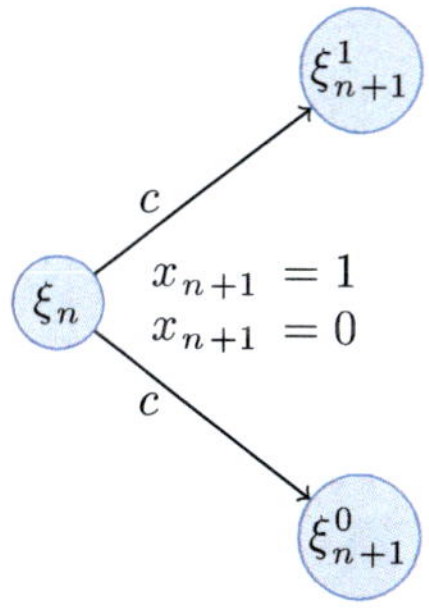

Fig. 5.4 A partial view of the multi-stage process

depending on what the value of the next observation x_n is. This is illustrated in Fig. 5.4, by extension from the previous two-stage example. The immediate value

$$V^0(\xi_t) = \sup_{a \in \mathcal{A}} \int_{\Theta} u(\theta, a) \, \mathrm{d}\xi_t(\theta)$$

is the expected value if we stop immediately at time t. The next-step value

$$\mathbb{E}_{\xi_n} V^0(\xi_{n+1}) = \int_{\mathcal{X}} V^0[\xi_n(\cdot \mid x_n)] \, \mathrm{d}\xi_n(x_n)$$

is the expected value of the next step, ignoring the cost. Finally, the optimal value at the n-th step is just the maximum of the value of stopping immediately and the next-step value, that is,

$$V^1(\xi_n) = \max \left\{ V^0(\xi_n), \mathbb{E}_{\xi_n} V^0(\xi_{n+1}) - c \right\}.$$

This procedure can be generalized over all steps $1, 2, \ldots, T$, to obtain a general procedure.

5.2.2 Backwards Induction for Bounded Procedures

The main idea expressed in the previous section is to start from the last stage of our decision problem, where the utility is known, and then move backwards. At each stage, we know the probability of reaching different points in the next stage, as well as their values. Consequently, we can compute the value of any point in the current stage as well. This idea is formalised below, via the algorithm of backwards induction.

Theorem 5.2.1 (Backwards induction) *The utility of a T-bounded optimal procedure with prior ξ_0 is $V^T(\xi_0)$ and is given by the recursion*

$$V^{j+1}(\xi_n) = \max \left\{ V^0(\xi_n), \mathbb{E}_{\xi_n} V^j(\xi_{n+1}) - c \right\} \tag{5.6}$$

for every belief ξ_n *in the set of beliefs that arise from the prior* ξ_0*, with* $j = T - n$.

The proof of this theorem follows by induction. However, we shall prove a more general version in Chap. 6. Equation 5.6 essentially gives a recursive calculation of the value of the T-bounded optimal procedure. To evaluate it, we first need to calculate all possible beliefs $\xi_1, \ldots, \xi_T$. For each belief ξ_T, we calculate $V^0(\xi_T)$. We then move backwards, and calculate $V^0(\xi_{T-1})$ and $V^1(\xi_{T-1})$. Proceeding backwards, for $n = T-1, T-2, \ldots, 1$, we calculate $V^{T+1}(\xi_n)$ for all beliefs ξ_n with $j = T - n$. The value of the procedure also determines the optimal sampling strategy, as shown by the following theorem.

Theorem 5.2.2 *The optimal* T*-bounded procedure stops at time* t *if the value of stopping at* t *is better than that of continuing, i.e. if*

$$V^0(\xi_t) \geq V^{T-t}(\xi_t).$$

This procedure chooses a maximizing $\mathbb{E}_{\xi_t} U(\theta, a)$*, otherwise takes one more sample.*

Finally, longer procedures (i.e. procedures that allow for stopping later) are always better than shorter ones, as shown by the following theorem.

Theorem 5.2.3 *For any probability measure* ξ *on* Θ*,*

$$V^n(\xi) \leq V^{n+1}(\xi). \tag{5.7}$$

That is, the procedure that stops after at most n steps is never better than the procedure that stops after at most $n+1$ time steps. To obtain an intuition of why this is the case, consider the example of Sect. 5.1.1. In that example, if we have a sequence of 3s, then we obtain no information. Consequently, when we compare the value of a plan taking at most n samples with that of a plan taking at most $n+1$ samples, we see that the latter plan is better for the event where we obtain n 3s, but has the same value for all other events.

5.2.3 Unbounded Sequential Decision Procedures

Given the monotonicity of the value of bounded procedures (5.7), one may well ask what is the value of unbounded procedures, i.e. procedures that may never stop sampling. The value of an unbounded sampling and decision procedure π under prior ξ is

$$U(\xi, \pi) = \int_{\mathcal{X}^*} \left\{V^0[\xi(\cdot \mid x^n)] - cn\right\} \mathrm{d}P_\xi^\pi(x^n) = \mathbb{E}_\xi^\pi \left\{V^0[\xi(\cdot \mid x^n)] - cn\right\},$$

where $P_\xi^\pi(x^n)$ is the probability that we observe samples x^n and stop under the marginal distribution defined by ξ and π, while n is the random number of samples taken by π. As before, this is random because the observations x are random; π itself can be deterministic.

Definition 5.2.2 (*Regular procedure*) Given a decision procedure π, let $B_{>k}(\pi) \subset \mathcal{X}^*$ be the set of sequences such that π takes more than k samples. Then π is regular if $U(\xi, \pi) \geq V^0(\xi)$ and if, for all $n \in \mathbb{N}$, and for all $x^n \in B_{>n}(\pi)$

$$U[\xi(\cdot \mid x^n), \pi] \geq V^0[\xi(\cdot \mid x^n)] - cn, \tag{5.8}$$

i.e., the expected utility given for any sample that starts with x^n where we don't stop, is greater than that of stopping at n.

In other words, if π specifies that at least one observation should be taken, then the value of π is greater than the value of choosing a decision without any observation. Furthermore, whenever π specifies that another observation should be taken, the expected value of continuing must be larger than the value of stopping. If the procedure is *not* regular, then there may be stages where the procedure specifies that sampling should be continued, though the value may not increase by doing so.

Theorem 5.2.4 *If π is not regular, then there exists a regular π' such that $U(\xi, \pi') \geq U(\xi, \pi)$.*

Proof First, consider the case that π is not regular because $U(\xi, \pi) \leq V^0(\xi)$. Then π' can be the regular procedure which chooses $a \in \mathcal{A}$ without any observations.

Now consider the case that $U(\xi, \pi) > V^0(\xi)$ and that π specifies at least one sample should be taken. Let π' be the procedure which stops as soon as the observed x^n does not satisfy (5.8).

If π stops, then both sides of (5.8) are equal, as the value of stopping immediately is at least as high as that of continuing. Consequently, π' stops no later than π for any x^n. Finally, let

$$B_k(\pi) = \left\{x \in \mathcal{X}^* \mid n = k\right\}$$

be the set of observations such that exactly k samples are taken by rule π. Given that π' does stop, we have

$$\begin{aligned} U(\xi, \pi') &= \sum_{k=1}^{\infty} \int_{B_k(\pi')} \{V^0[\xi(\cdot \mid x^k) - ck]\} \, \mathrm{d}P_\xi(x^k) \\ &\geq \sum_{k=1}^{\infty} \int_{B_k(\pi')} U[\xi(\cdot \mid x^k, \pi)] \, \mathrm{d}P_\xi(x^k) \\ &= \sum_{k=1}^{\infty} \mathbb{E}_\xi^\pi\{U \mid B_k(\pi')\} P_\xi(B_k(\pi')) = E_\xi^\pi U = U(\xi, \pi), \end{aligned}$$

where the inequality follows from the fact that π is not regular. □

5.2.4 The Sequential Probability Ratio Test

Sometimes we wish to collect just enough data in order to be able to confirm or disprove a particular hypothesis. More specifically, we have a set of parameters Θ, and we need to pick the right one. However, rather than simply using an existing set of data, we are collecting data sequentially, and we need to decide when to stop and select a model. In this case, each one of our decisions a_i corresponds to choosing the model θ_i, and we have a utility function that favours our picking the correct model. As before, data collection has some cost, which we must balance against the expected utility of picking a parameter.

As an illustration, consider a problem where we must decide for one out of two possible parameters θ_1, θ_2. At each step, we can either take another sample from the unknown $P_\theta(x_t)$, or decide for one or the other of the parameters.

Example 5.3 *(A two-point sequential decision problem.)* Consider a problem where there are two parameters and two final actions which select one of the two parameter values, such that:

- Observations $x_t \in \mathcal{X}$
- Distribution family: $\mathscr{P} = \{P_\theta \mid \theta \in \Theta\}$
- Probability space $(\mathcal{X}^*, \mathfrak{B}(\mathcal{X}^*), P_\theta)$.
- Parameter set $\Theta = \{\theta_1, \theta_2\}$.
- Action set $\mathcal{A} = \{a_1, a_2\}$.
- Prior $\xi = \mathbb{P}(\theta = \theta_1)$.
- Sampling cost $c > 0$.

The actions we take upon stopping can be interpreted as guessing the parameter. When we guess wrong, we suffer a cost, as seen in Table 5.1.

As will be the case for all our sequential decision problems, we only need to consider our current belief ξ, and its possible evolution, when making a decision. To obtain some intuition about this procedure, we are going to analyse this problem by examining what the optimal decision is under all possible beliefs ξ.

Under some belief ξ, the immediate value (i.e. the value we obtain if we stop immediately) is simply

$$V^0(\xi) = \max\{\lambda_1 \xi, \lambda_2(1 - \xi)\}. \tag{5.9}$$

The worst-case immediate value, i.e. the minimum, is attained when both terms are equal. Consequently, setting $\lambda_1 \xi = \lambda_2(1 - \xi)$ gives $\xi = \lambda_2/(\lambda_1 + \lambda_2)$. Intuitively,

Table 5.1 The local utility function, with $\lambda_1, \lambda_2 < 0$

$U(\theta, d)$	a_1	a_2
θ_1	0	λ_1
θ_2	λ_2	0

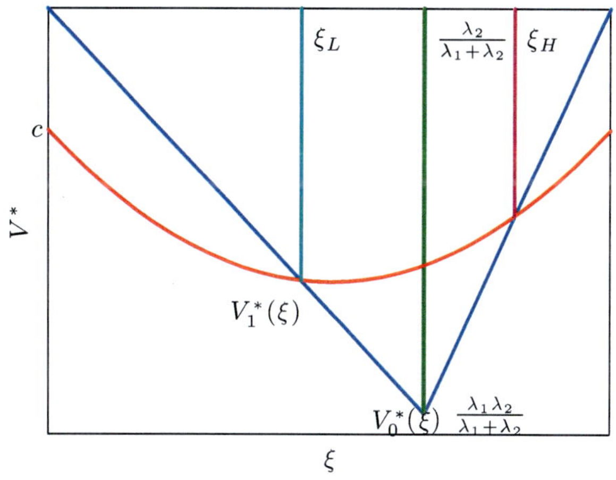

Fig. 5.5 The value of the optimal continuation V' versus stopping V^0

this is the worst-case belief, as the uncertainty it induces leaves us unable to choose between either hypothesis. Replacing in (5.9) gives a lower bound for the value for any belief:

$$V^0(\xi) \geq \frac{\lambda_1\lambda_2}{\lambda_1+\lambda_2}.$$

Let Π denote the set of procedures π which take at least one observation and define

$$V'(\xi) = \sup_{\pi \in \Pi} U(\xi, \pi).$$

Then the ξ-expected utility $V^*(\xi)$ must satisfy

$$V^*(\xi) = \max\left\{V^0(\xi), V'(\xi)\right\}.$$

As we showed in Sect. 3.3.1, V' is a convex function of ξ. Now let

$$\Xi_0 \triangleq \left\{\xi \mid V^0(\xi) \geq V'(\xi)\right\}$$

be the set of priors where it is optimal to terminate sampling. It follows that $\Xi \setminus \Xi_0$, the set of priors where we must not terminate sampling, is a convex set.

Figure 5.5 illustrates the above arguments, by plotting the immediate value against the optimal continuation after taking one more sample. For the worst-case belief, we must always continue sampling. When we are absolutely certain about the model, then it's always better to stop immediately. There are two points where the curves intersect. Together, these define three subsets of beliefs: On the left, if $\xi < \xi_L$, we decide for one parameter θ_1. On the right, if $\xi > \xi_H$, we decide for the other parameter, θ_2. Otherwise, we continue sampling. This is the main idea of the sequential probability ratio test, explained below.

The sequential probability ratio test (SPRT)
Figure 5.5 offers a graphical illustration of when it is better to take one more sample in this setting. In particular, if $\xi \in (\xi_L, \xi_T)$, then it is optimal to take at least one more sample. Otherwise, it is optimal to make an immediate decision with value $\rho_0(\xi)$.

This has a nice interpretation as a standard tool in statistics: the *sequential probability ratio test.* First note that our posterior at time t can be written as

$$\xi_t = \frac{\xi P_{\theta_1}(x^t)}{\xi P_{\theta_1}(x^t) + (1-\xi) P_{\theta_2}(x^t)}.$$

Then, for any posterior, the optimal procedure is:

- If $\xi_L < \xi_t < \xi_T$, take one more sample.
- If $\xi_L \geq \xi_t$, stop and choose a_2.
- If $\xi_T \leq \xi_t$, stop and choose a_1.

We can now restate the optimal procedure in terms of a probability ratio, i.e. we should always take another observation as long as

$$\frac{\xi(1-\xi_T)}{(1-\xi)\xi_T} < \frac{P_{\theta_2}(x^t)}{P_{\theta_1}(x^t)} < \frac{\xi(1-\xi_L)}{(1-\xi)\xi_L}.$$

If the first inequality is violated, we choose a_1. If the second inequality is violated, we choose a_2. So, there is an equivalence between SPRT and optimal sampling procedures, when the optimal policy is to continue sampling whenever our belief is within a specific interval.

5.2.5 *Wald's Theorem*

An important tool in the analysis of SPRT as well as other procedures that stop at random times is the following theorem by Wald.

Theorem 5.2.5 (Wald's theorem) *Let $z_1, z_2, \ldots$ be a sequence of i.i.d. random variables with measure G, such that $\mathbb{E}\, z_i = m$ for all i. Then for any sequential procedure with $\mathbb{E}\, n < \infty$:*

$$\mathbb{E} \sum_{i=1}^{n} z_i = m\, \mathbb{E}\, n. \tag{5.10}$$

Proof In the following proof, we omit explicit references to the implied sequential procedure.

$$
\begin{aligned}
\mathbb{E}\sum_{i=1}^{n} z_i &= \sum_{k=1}^{\infty}\int_{B_k}\sum_{i=1}^{k} z_i \, \mathrm{d}G^k(z^k) \\
&= \sum_{k=1}^{\infty}\sum_{i=1}^{k}\int_{B_k} z_i \, \mathrm{d}G^k(z^k) \\
&= \sum_{i=1}^{\infty}\sum_{k=i}^{\infty}\int_{B_k} z_i \, \mathrm{d}G^k(z^k) \\
&= \sum_{i=1}^{\infty}\int_{B_{\geq i}} z_i \, \mathrm{d}G^i(z^i) \\
&= \sum_{i=1}^{\infty}\mathbb{E}(z_i)\,\mathbb{P}(n \geq i) = m\,\mathbb{E}\,n
\end{aligned}
$$

□

We now consider an application of this theorem to the SPRT. Let $z_i = \log \frac{P_{\theta_2}(x_i)}{P_{\theta_1}(x_i)}$. Consider the equivalent formulation of the SPRT which uses

$$a < \sum_{i=1}^{n} z_i < b$$

as the test. Using Wald's theorem and the previous properties and assuming $c \approx 0$, we obtain the following approximately optimal values for a, b:

$$a \approx \log c - \log \frac{I_1 \lambda_2 (1-\xi)}{\xi}, \qquad b \approx \log \frac{1}{c} - \log \frac{I_2 \lambda_1 \xi}{1-\xi},$$

where $I_1 = -\,\mathbb{E}(z \mid \theta = \theta_1)$ and $I_2 = \mathbb{E}(z \mid \theta = \theta_2)$ is the *information*, better known as the *KL-divergence*. If the cost c is very small, then the information terms vanish and we can approximate the values by $\log c$ and $\log \frac{1}{c}$.

5.3 Martingales

Martingales are a fundamentally important concept in the analysis of stochastic processes where the expectation at time $t + 1$ only depends on the state of the process at time t.

An example of a martingale sequence is when x_t is the amount of money you have at a given time, and where at each time-step t you are making a gamble such that you lose or gain 1 currency unit with equal probability. Then, at any step t, it holds that $\mathbb{E}(x_{t+1} \mid x_t) = x_t$. This concept can be generalized to two random processes x_t and y_t, which are dependent.

Definition 5.3.1 Let $x^n \in \mathcal{S}^n$ be a sequence of observations with distribution P_n, and $y_n : \mathcal{S}^n \to \mathbb{R}$ be a random variable. Then the sequence $\{y_n\}$ is a *martingale with* *respect to* $\{x_n\}$ if for all n the expectation

$$\mathbb{E}(y_n) = \int_{\mathcal{S}^n} y_n(x^n) \, \mathrm{d} P_n(x^n)$$

exists and

$$\mathbb{E}(y_{n+1} \mid x^n) = y_n$$

holds with probability 1. If $\{y_n\}$ is a martingale with respect to itself, i.e. $y_i(x) = x$, then we call it simply a *martingale*.

It is also useful to consider the following generalizations of martingale sequences.

Definition 5.3.2 A sequence $\{y_n\}$ is a super-martingale if $\mathbb{E}(y_{n+1} \mid x^n) \leq y_n$ and a sub-martingale if $\mathbb{E}(y_{n+1} \mid x^n) \geq y_n$, w.p. 1.

At a first glance, it might appear that martingales are not very frequently encountered, apart from some niche applications. However, we can always construct a martingale from any sequence of random variables as follows.

Definition 5.3.3 (*Doob martingale*) Consider a function $f : \mathcal{S}^m \to \mathbb{R}$ and some associated random variables $x^m \triangleq x_1, \ldots x_m$. Then, for any $n \leq m$, assuming the expectation $\mathbb{E}(f \mid x^n) = \int_{\mathcal{S}^{m-n}} f(x^m) \, \mathrm{d} \, \mathbb{P}(x_{n+1}, \ldots, x_m \mid x^n)$ exists, we can construct the random variable

$$y_n(x^n) = \mathbb{E}[f \mid x^n].$$

Then $\mathbb{E}(y_{n+1} \mid x^n) = y_n$, and so y_n is a martingale sequence with respect to x_n.

Another interesting type of martginale sequence are martingale *difference* sequences. They are particularly important as they are related to some useful concentration bounds.

Definition 5.3.4 A sequence $\{y_n\}$ is a *martingale difference sequence* with respect to $\{x_n\}$ if

$$\mathbb{E}(y_{n+1} \mid x^n) = 0 \qquad \text{with probability 1.}$$

For bounded difference sequences, the following well-known concentration bound holds.

Theorem 5.3.1 *Let b_k be a random variable depending on x^{k-1} and $\{y_k\}$ be a martingale difference sequence with respect to the $\{x_k\}$, such that $y_k \in [b_k, b_k + c_k]$ w.p. 1, Then, defining $s_k \triangleq \sum_{i=1}^{k} y_i$, it holds that*

$$\mathbb{P}(s_n \geq t) \leq \exp\left(\frac{-2t^2}{\sum_{i=1}^{n} c_i^2}\right). \tag{5.11}$$

This allows us to bound the probability that the difference sequence deviates from zero. Since there are only few problems where the default random variables are difference sequences, use of this theorem is most common by defining a new random variable sequence that is a difference sequence.

5.4 Markov Processes

A more general type of sequence of random variables than martingales are Markov processes. Informally speaking, a Markov process is a sequence of variables $\{x_n\}$ such that the next value x_{t+1} only depends on the current value x_t.

Definition 5.4.1 (*Markov Process*) Let $(\mathcal{S}, \mathfrak{B}(\mathcal{S}))$ be a measurable space. If $\{x_n\}$ is a sequence of random variables $x_n : \mathcal{S} \to \mathcal{X}$ such that

$$\mathbb{P}(x_t \in A \mid x_{t-1}, \dots, x_1) = \mathbb{P}(x_t \in A \mid x_{t-1}), \qquad \forall A \in \mathfrak{B}(\mathcal{X}),$$

i.e., x_t is independent of $x_{t-2}, \dots$ given x_{t-1}, then $\{x_n\}$ is a Markov process, and x_t is called the *state* of the Markov process at time t. If $\mathbb{P}(x_t \in A \mid x_{t-1} = x) = \tau(A \mid x)$ where $\tau : \mathfrak{B}(\mathcal{S}) \times \mathcal{S} \to [0, 1]$ is the *transition kernel*, then $\{x_n\}$ is a *stationary Markov process*

stationary Markov process

Note that is the sequence of posterior parameters obtained in Bayesian inference is a Markov process.

5.5 Exercises

Exercise 5.1 Consider a stationary Markov process with state space S and whose transition kernel is a matrix $\boldsymbol{\tau}$. At time t, we are at state $x_t = s$ and we can either, 1: Terminate and receive reward $b(s)$, or 2: Pay $c(s)$ and continue to a random state x_{t+1} from the distribution $\boldsymbol{\tau}(z' \mid z)$.

Assuming $b, c > 0$ and $\boldsymbol{\tau}$ are known, design a backwards induction algorithm that optimizes the utility function

$$U(x_1, \dots, x_T) = b(x_T) - \sum_{t=1}^{T-1} c(x_t).$$

Finally, show that the expected utility of the optimal policy starting from any state must be bounded.

Exercise 5.2 Consider the problem of classification with features $x \in \mathcal{X}$ and labels $y \in \mathcal{Y}$, where each label costs $c > 0$. Assume a Bayesian model with some parameter space Θ on which we have a prior distribution ξ_0. Let ξ_t be the posterior distribution after t examples $(x_1, y_1), \dots, (x_t, y_t)$.

Let our expected utility be the expected accuracy (i.e., the marginal probability of correctly guessing the right label over all possible models) of the Bayes-optimal classifier $\pi : \mathcal{X} \to \mathcal{Y}$, minus the cost paid, i.e.,

$$\mathbb{E}_t(U) \triangleq \max_{\pi} \int_{\Theta} \int_{\mathcal{X}} P_\theta(\pi(x) \mid x) \, \mathrm{d}P_\theta(x) \, \mathrm{d}\xi_t(\theta) - ct.$$

Show that the Bayes-optimal classification accuracy (ignoring the label cost) after t observations can be rewritten as

$$\int_{\Theta} \int_{\mathcal{X}} \max_{y \in \mathcal{Y}} P_\theta(y \mid x) \, \mathrm{d}\xi_t(\theta \mid x) \, \mathrm{d}\,\mathbb{P}_t(x),$$

where $\mathbb{P}_t$ and $\mathbb{E}_t$ denote marginal distributions under the belief ξ_t. Write the expression for the expected gain in accuracy when obtaining one more sample and label.

Implement the above for a model family of your choice. Two simple options are the following. The first is a finite model family composed of two different classifiers $P_\theta(y \mid x)$. The second is the family of discrete classifier models with a Dirichlet product prior, i.e. where $\mathcal{X} = \{1, \dots, n\}$, and each different $x \in \mathcal{X}$ corresponds to a different multinomial distribution over $\mathcal{Y}$. In both cases, you can assume a common (and known) data distribution $P(x)$, in which case $\xi_t(\theta \mid x) = \xi_t(\theta)$.

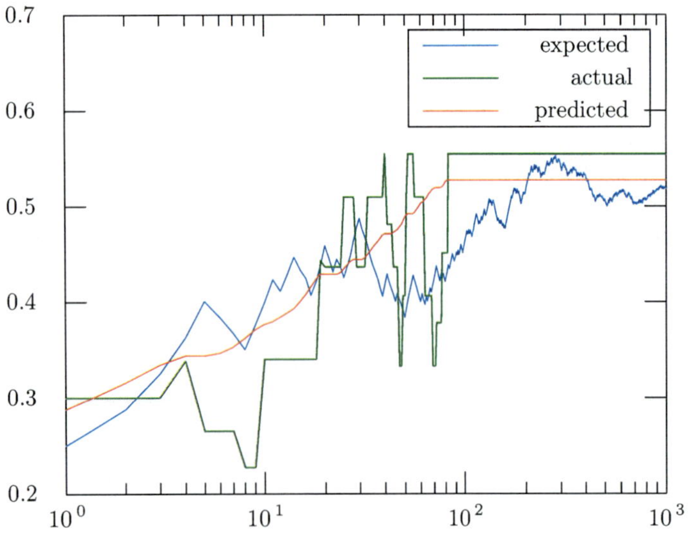

Fig. 5.6 Illustrative results for an implementation of Exercise 5.2 on a discrete classifier model

Figure 5.6 shows the performance for a family of discrete classifier models with $|\mathcal{X}| = 4$. It shows the **expected** classifier performance (based on the posterior marginal), the **actual** performance on a small test set, as well as the cumulative **predicted** performance gain. As you can see, even though the expected performance gain is zero in some cases, cumulatively it reaches the actual performance of the classifier. You should be able to produce a similar figure for your own setup.

Chapter 6
Experiment Design and Markov Decision Processes

6.1 Introduction

This chapter introduces the very general formalism of *Markov decision processes* (*MDPs*) that allows representation of various sequential decision making problems. Thus a Markov decision process can be used to model stochastic path problems, stopping problems as well as problems in reinforcement learning, experiment design, and control.

We begin by taking a look at the problem of *experimental design*. A typical question is to how to best allocate treatments with unknown efficacy to patients in an adaptive manner, so that the best treatment is found, or so as to maximize the number of patients that are treated successfully. The problem, originally considered by Chernoff [1, 2], informally can be stated as follows. *experimental design*

We have a number of treatments of unknown efficacy, i.e., some of them work better than the others. We observe patients one at a time. When a new patient arrives, we must choose which treatment to administer. Afterwards, we observe whether the patient improves or not. Given that the treatment effects are initially unknown, how can we maximize the number of cured patients? Alternatively, how can we discover the best treatment? The two different problems are formalized below.

Example 6.1 [*Adaptive treatment allocation*] Consider K treatments to be administered to T volunteers. To each volunteer only a single treatment can be assigned. At the tth trial, we treat one volunteer with some treatment $a_t \in \{1, \ldots, K\}$. We then obtain a reward $r_t = 1$ if the patient is healed and 0 otherwise. We wish to choose treatments maximizing the utility $U = \sum_t r_t$. This corresponds to maximizing the number of patients that get healed over time. *adaptive treatment allocation*

Example 6.2 [*Adaptive hypothesis testing*] An alternative goal would be to do a *clinical trial*, in order to find the best possible treatment. For simplicity, consider the problem of trying to find out whether a particular treatment is better or not than a placebo. We are given a hypothesis set Ω, with each $\omega \in \Omega$ corresponding to different *adaptive hypothesis testing* *clinical trial*

C. Dimitrakakis and R. Ortner, *Decision Making Under Uncertainty and Reinforcement Learning*, Intelligent Systems Reference Library 223,
https://doi.org/10.1007/978-3-031-07614-5_6

models for the effect of the treatment and the placebo. Since we don't know what is the right model, we place a prior ξ_0 on Ω. We can perform T experiments, after which we must decide whether or not the treatment is significantly better than the placebo. To model this, we define a decision set $\mathcal{D} = \{d_0, d_1\}$ and a utility function $U : \mathcal{D} \times \Omega \to \mathbb{R}$ which models the effect of each decision d given different versions of reality ω. One hypothesis $\omega \in \Omega$ is true. To identify the correct hypothesis, we can choose from a set of K possible experiments to be performed over T trials. At the t-th trial, we choose experiment $a_t \in \{1, \dots, K\}$ and observe outcome $x_t \in \mathcal{X}$, with $x_t \sim P_\omega$ drawn from the true hypothesis. Our posterior is

$$\xi_t(\omega) \triangleq \xi_0(\omega \mid a_1, \dots, a_t, x_1, \dots, x_t).$$

The reward is $r_t = 0$ for $t < T$ and

$$r_T = \max_{d \in D} \mathbb{E}_{\xi_T}(U \mid d).$$

Our utility can again be expressed as a sum over individual rewards, $U = \sum_{t=1}^{T} r_t$.

Both formalizations correspond to so-called *bandit problems* which we take a closer look at in the following section.

6.2 Bandit Problems

The simplest bandit problem is the stochastic n-armed bandit. We are faced with K different one-armed bandit machines, such as those found in casinos. At time t, we have to choose one *action* (i.e., a machine, in the bandit context usually termed *arm*) $a_t \in \mathcal{A} = \{1, \dots, K\}$. Each time t we play a machine, we receive a reward r_t with fixed expected value $\omega_i = \mathbb{E}(r_t \mid a_t = i)$. Unfortunately, we do not know the ω_i, and consequently the best arm is also unknown. The question is how to choose arms in order to maximize the total expected reward.

Definition 6.2.1 (*The stochastic n-armed bandit problem*) Given a set of arms $\mathcal{A} = \{1, \dots, K\}$ each giving reward according to a fixed reward distribution with unknown mean, select a sequence of actions $a_t \in \mathcal{A}$, so as to maximize expected utility, where the utility is

$$U = \sum_{t=0}^{T-1} \gamma^t r_t,$$

discount factor

$T \in (0, \infty]$ is the *horizon*, and $\gamma \in (0, 1]$ a *discount factor*. The reward r_t is stochastic and only depends on the action chosen at step t with expectation $\mathbb{E}(r_t \mid a_t = i) = \omega_i$.

policy

For selecting actions, we want to specify some *policy* or decision rule. Such a rule shall only depend on the sequence of previously taken actions and observed rewards.

Usually, the policy $\pi : \mathcal{A}^* \times \mathbb{R}^* \to \mathcal{A}$ is a deterministic mapping from the space of all sequences of actions and rewards to actions. That is, for every observation and action history $a_1, \dots, a_{t-1}, r_1, \dots, r_{t-1}$ it suggests a single action a_t. More generally, it could also be a stochastic policy, that specifies a mapping to action distributions. We use the notation

$$\pi(a_t \mid a^{t-1}, r^{t-1})$$

for stochastic history-dependent bandit policies, i.e., the probability of choosing action a_t given the history until time t.

One idea to approach the bandit problem is to apply the Bayesian decision-theoretic framework we have developed earlier to maximize utility in expectation. More specifically, given the horizon $T \in (0, \infty]$ and the discount factor $\gamma \in (0, 1]$, we define our utility from time t to be

$$U_t \triangleq \sum_{k=1}^{T-t} \gamma^k r_{t+k}.$$

As a first step we need to define a suitable family of probability measures $\mathscr{P}$, indexed by parameter $\omega \in \Omega$ describing the reward distribution of each bandit, together with a prior distribution ξ on Ω. Since ω is unknown, we cannot maximize the expected utility with respect to it. However, we can always maximize expected utility with respect to our belief ξ. That is, we replace the ill-defined problem of maximizing utility in an unknown model with that of maximizing expected utility given a distribution over possible models. The problem can be written in a simple form as

$$\max_{\pi} \mathbb{E}_{\xi}^{\pi} U_t = \max_{\pi} \int_{\Omega} \mathbb{E}_{\omega}^{\pi} U_t \, d\xi(\omega). \tag{6.2.1}$$

The following figure summarizes the bandit problem formulated in the Bayesian setting.

Decision-theoretic statement of the bandit problem

- Let $\mathcal{A}$ be the set of arms.
- Define a family of distributions $\mathscr{P} = \left\{P_{\omega,i} \mid \omega \in \Omega, i \in \mathcal{A}\right\}$ on $\mathbb{R}$.
- Assume the i.i.d. model $r_t \mid \omega, a_t = i \sim P_{\omega,i}$.
- Dcfinc prior ζ on Ω.
- Select a policy $\pi : \mathcal{A}^* \times \mathbb{R}^* \to \mathcal{A}$ maximizing

$$\mathbb{E}_{\xi}^{\pi} U = \mathbb{E}_{\xi}^{\pi} \sum_{t=0}^{T-1} \gamma^t r_t.$$

There are two main difficulties with this approach. The first one is that the choice of the family and the prior distribution is effectively part of the problem formulation and can severely influence the solution. This issue can be resolved by either specifying a subjective prior distribution, or by selecting a prior distribution that has good worst-case guarantees. The second difficulty concerns the computation of the policy that maximizes expected utility given a prior and a family. This is a hard problem, because in general such policies are history dependent and the set of all possible histories is exponential in the horizon T.

6.2.1 An Example: Bernoulli Bandits

As a simple illustration, consider the case when the reward for choosing one of the n actions is either 0 or 1 with some fixed yet unknown probability depending on the chosen action. This can be modelled in the standard Bayesian framework using the Beta-Bernoulli conjugate prior. More specifically, we can formalize the problem as follows.

Consider n Bernoulli distributions with unknown parameters ω_i $(i = 1, \dots, n)$ such that

$$r_t \mid a_t = i \sim \mathcal{Bern}(\omega_i), \qquad \mathbb{E}(r_t \mid a_t = i) = \omega_i. \tag{6.2.2}$$

Each Bernoulli distribution thus corresponds to the distribution of rewards obtained from each bandit that we can play. In order to apply the statistical decision theoretic framework, we have to quantify our uncertainty about the parameters ω in terms of a probability distribution.

We model our belief for each bandit's parameter ω_i as a Beta distribution $\mathcal{Beta}(\alpha_i, \beta_i)$, with density $f(\omega \mid \alpha_i, \beta_i)$ so that

$$\xi(\omega_1, \dots, \omega_n) = \prod_{i=1}^{n} f(\omega_i \mid \alpha_i, \beta_i).$$

Recall that the posterior of a Beta prior is also a Beta. Let

$$N_{t,i} \triangleq \sum_{k=1}^{t} \mathbb{I}\{a_k = i\}$$

be the number of times we played arm i and

$$\hat{r}_{t,i} \triangleq \frac{1}{N_{t,i}} \sum_{k=1}^{t} r_t \cdot \mathbb{I}\{a_k = i\}$$

be the *empirical reward* of arm i at time t. We can set $\hat{r}_{t,i} = 0$ when $N_{t,i} = 0$. Then, the posterior distribution for the parameter of arm i is

$$\xi_t = \mathcal{B}eta(\alpha_i + N_{t,i}\,\hat{r}_{t,i}\,,\ \beta_i + N_{t,i}\,(1 - \hat{r}_{t,i})).$$

In order to evaluate a policy we need to be able to predict the expected utility we obtain. The latter only depends on our current belief, and the state of our belief corresponds to the state of the bandit problem. This means that everything we know about the problem at time t can be summarized by ξ_t. For Bernoulli bandits, a sufficient statistic for our belief is the number of times we played each bandit and the total reward from each bandit. Thus, our state at time t is entirely described by our priors α, β (the initial state) and the vectors

$$N_t = (N_{t,1}, \ldots, N_{t,n})$$
$$\hat{r}_t = (\hat{r}_{t,1}, \ldots, \hat{r}_{t,n}).$$

Accordingly, as $r_t \in \{0, 1\}$, the possible states of our belief given some prior are in $\mathbb{N}^{2n}$. At any time t, we can calculate the probability of observing $r_t = 1$ if we pull arm i as

$$\xi_t(r_t = 1 \mid a_t = i) = \frac{\alpha_i + N_{t,i}\,\hat{r}_{t,i}}{\alpha_i + \beta_i + N_{t,i}}.$$

So, not only we can predict the immediate reward based on our current belief, but we can also predict all possible next beliefs: the next state is well-defined and depends only on the current state and observation. As we shall see later, this type of decision problem can be modelled as a Markov decision process (Definition 6.3.1). For now, we shall take a closer look at the general bandit process itself.

6.2.2 Decision-Theoretic Bandit Process

The basic view of the bandit process is to consider only the decision maker's actions a_t, obtained rewards r_t and the latent parameter ω, as shown in Fig. 6.2a. With this basic framework, we can now define the general decision-theoretic bandit process, which also includes the states of belief ξ_t of the decision maker.

Definition 6.2.2 Let $\mathcal{A}$ be a set of arms (now not necessarily finite). Let Ω be a set of possible parameter values, indexing a family of probability measures $\mathscr{P} = \{P_{\omega,a} \mid \omega \in \Omega, a \in \mathcal{A}\}$. There is some $\omega \in \Omega$ such that, whenever we take action $a_t = i$, we observe reward $r_t \in \mathcal{R} \subset \mathbb{R}$ with probability measure

$$P_{\omega,i}(R) \triangleq \mathbb{P}_\omega(r_t \in R \mid a_t = i), \qquad R \subseteq \mathcal{R}.$$

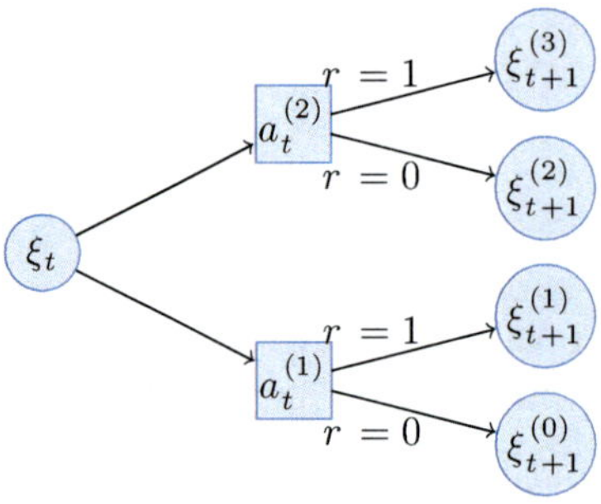

Fig. 6.1 A partial view of the multi-stage process. Here, the probability that we obtain $r = 1$ if we take action $a_t = i$ is simply $P_{\xi_t,i}(\{1\})$

Let ξ_0 be a prior distribution on Ω and let the posterior distributions for $B \subseteq \Omega$ be defined as

$$\xi_{t+1}(B) \propto \int_B P_{\omega,a_t}(r_t) \, \mathrm{d}\xi_t(\omega).$$

The next belief ξ_{t+1} is random, since it depends on the random quantity r_t. In fact, the probability of the next reward lying in some set R for $a_t = i$ is given by the marginal distribution

$$P_{\xi_t,i}(R) \triangleq \int_\Omega P_{\omega,i}(R) \, \mathrm{d}\xi_t(\omega).$$

Finally, as ξ_{t+1} deterministically depends on ξ_t, a_t, r_t, the probability of obtaining a particular next belief is the same as the probability of obtaining the corresponding rewards leading to the next belief. In more detail, we can write

$$\mathbb{P}(\xi_{t+1} = \xi \mid \xi_t, a_t) = \int_{\mathcal{R}} \mathbb{I}\ \xi_t(\cdot \mid a_t, r_t = r) = \xi \ \mathrm{d}P_{\xi_t,a}(r).$$

In practice, although multiple reward sequences may lead to the same beliefs, we frequently ignore that possibility for simplicity. Then the process obtains a tree-like structure. A solution to the problem of which action to select is given by the following *backwards induction* algorithm for bandits, similar to the backwards induction algorithm given in Sect. 5.2.2, i.e.,

backwards induction

$$U^*(\xi_t) = \max_{a_t} \mathbb{E}(r_t \mid \xi_t, a_t) + \sum_{\xi_{t+1}} \mathbb{P}(\xi_{t+1} \mid \xi_t, a_t) \, U^*(\xi_{t+1}).$$

If you look at this structure, you can see that the next belief only depends on the current belief, action, and reward, i.e., it satisfies the Markov property, as seen in Fig. 6.1.

In reality, the reward depends only on the action and the unknown ω, as can be seen in Fig. 6.2a. This is the point of view of an external observer. If we want to

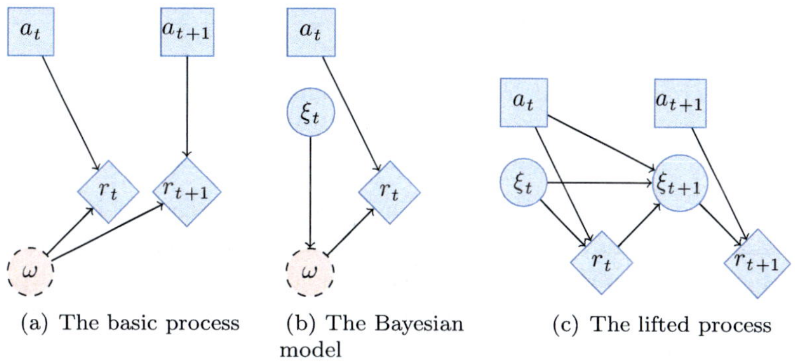

Fig. 6.2 Three views of the bandit process. The figure (**a**) shows the basic bandit process from the view of an external observer. The decision maker selects a_t and then obtains reward r_t, while the parameter ω is hidden. The process is repeated for $t = 1, \ldots, T$. The Bayesian model is shown in (**b**) and the resulting process in (**c**). While ω is not known, at each time step t we maintain a belief ξ_t on Ω. The reward distribution is then defined through our belief. In (**b**), we can see the complete process, where the dependency on ω is clear. In (**c**), we marginalize out ω and obtain a model where the transitions only depend on the current belief and action

add the decision maker's internal belief to the graph, we obtain Fig. 6.2b. From the point of view of the decision maker, the distribution of ω only depends on his current belief. Consequently, the distribution of rewards also only depends on the current belief, as we can marginalize over ω. This gives rise to the decision-theoretic bandit process shown in Fig. 6.2c.

A decision-theoretic bandit process can be modelled more generally as a Markov decision process, a setting we shall consider more generally in the following section. It turns out that backwards induction, as well as other efficient algorithms, can provide optimal solutions for Markov decision processes, too.

6.3 Markov Decision Processes and Reinforcement Learning

The bandit setting is one of the simplest instances of so-called *reinforcement learning* problems. Informally, speaking, these are problems of learning how to act in an unknown environment, only through interaction with the environment and limited reinforcement signals. The learning agent interacts with the environment by choosing actions that give certain observations and rewards. The goal is usually to maximize some measure of the accumulated reward. For example, we can consider a mouse running through a maze, where the reward is finding one or all pieces of cheese hidden in the maze.

> **The reinforcement learning problem**
> *Learn* how to act in an *unknown* environment, only by *interaction* and *reinforcement.*

Generally, we assume that the environment μ that we are acting in has an underlying state $s_t \in \mathcal{S}$, which changes in discrete time steps t. At each step, the agent obtains an observation $x_t \in \mathcal{X}$ and chooses an action $a_t \in \mathcal{A}$. We usually assume that the environment is such that its next state s_{t+1} only depends on its current state s_t and the last action a_t taken by the agent. In addition, the agent observes a reward signal r_t, and its goal is to maximize the total reward during its lifetime.

When the environment μ is unknown, this problem is hard even in seemingly simple settings like the multi-armed bandit, where the underlying state never changes. In many real-world applications, the problem is even harder, as the state often is not directly observed. Instead, we may have to rely on the observables x_t, which give only partial information about the true underlying state s_t.

Reinforcement learning problems typically fall into one of the following three categories: (1) Markov decision processes (MDPs), where the state s_t is observed directly, i.e., $x_t = s_t$; (2) partially observable MDPs (POMDPs), where the state is hidden, i.e., x_t is only probabilistically dependent on the state s_t; and (3) stochastic Markov games, where the next state also depends on the move of other agents. While all of these problem descriptions are different, in the Bayesian setting they all can be reformulated as MDPs by constructing an appropriate belief state, similarly to the decision theoretic formulation of the bandit problem.

In this chapter, we shall confine our attention to Markov decision processes. Hence, we shall not discuss the case where we cannot observe the state directly, or consider the existence of other agents.

transition distribution

reward distribution

Definition 6.3.1 (*Markov decision process*) A Markov decision process (MDP) μ is a tuple $\mu = \langle \mathcal{S}, \mathcal{A}, \mathscr{P}, \mathscr{R} \rangle$, where $\mathcal{S}$ is the *state space* and $\mathcal{A}$ is the *action space*. The *transition distribution* $\mathscr{P} = \{p(\cdot \mid s, a) \mid s \in \mathcal{S}, a \in \mathcal{A}\}$ is a collection of probability measures on $\mathcal{S}$, indexed in $\mathcal{S} \times \mathcal{A}$, and the *reward distribution* $\mathscr{R} = \{\rho(\cdot \mid s, a) \mid s \in \mathcal{S}, a \in \mathcal{A}\}$ is a collection of probability measures on $\mathbb{R}$, such that

$$p(S \mid s, a) = \mathbb{P}_\mu(s_{t+1} \in S \mid s_t = s, a_t = a),$$
$$\rho(R \mid s, a) = \mathbb{P}_\mu(r_t \in R \mid s_t = s, a_t = a).$$

Usually, also an initial state s_0 (or more generally, an initial distribution from which s_0 is sampled) is specified.

In the following, we usually assume only MDPs with finite state and action spaces $\mathcal{S}, \mathcal{A}$. By definition, rewards and transition probabilities of MDP are time-invariant, although one could assume more general models. In any case however, rewards and transitions in an MDP μ shall satisfy the following Markov property (Fig. 6.3).

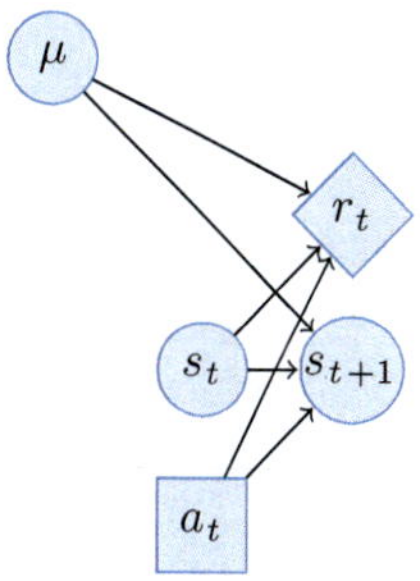

Fig. 6.3 The structure of a Markov decision process

Markov property of the reward and state distribution

$$\mathbb{P}_\mu(s_{t+1} \in S \mid s_1, a_1, \dots, s_t, a_t) = \mathbb{P}_\mu(s_{t+1} \in S \mid s_t, a_t), \quad \text{(Transition distribution)}$$

$$\mathbb{P}_\mu(r_t \in R \mid s_1, a_1, \dots, s_t, a_t) = \mathbb{P}_\mu(r_t \in R \mid s_t, a_t), \quad \text{(Reward distribution)}$$

where $S \subset \mathcal{S}$ and $R \subset \mathcal{R}$ are reward and state subsets, respectively.

Dependencies of rewards. In the following, for a given MDP μ we use

$$r_\mu(s, a) = \mathbb{E}_\mu(r_{t+1} \mid s_t = s, a_t = a)$$

to denote the expected reward for a state-action-pair s, a. Similarly, we write $p_\mu(\cdot \mid s, a)$ for the transition probability distribution under s, a.

Sometimes however it is more convenient to have rewards that depend on the next state as well, i.e.,

$$r_\mu(s, a, s') = \mathbb{E}_\mu(r_{t+1} \mid s_t = s, a_t = a, s_{t+1} = s'),$$

though this is complicates the notation considerably since now the reward is obtained on the next time step. However, we can always replace this with the expected reward for a given state-action pair, i.e.,

$$r_\mu(s, a) = \mathbb{E}_\mu(r_{t+1} \mid s_t = s, a_t = s) = \sum_{s' \in \mathcal{S}} p_\mu(s' \mid s, a) \, r_\mu(s, a, s').$$

In a simpler setting, rewards only depend on the current state, so that we can write

$$r_\mu(s) = \mathbb{E}_\mu(r_t \mid s_t = s).$$

policy

Policies. A *policy* π (sometimes also called *decision function*) specifies which action to take. One can think of a policy as implemented through an algorithm or an embodied agent.

Policies
A policy π defines a conditional distribution on actions given the history:

$$\mathbb{P}^\pi(a_t \mid s_t, \ldots, s_1, a_{t-1}, \ldots, a_1) \qquad \text{(history-dependent policy)}$$
$$\mathbb{P}^\pi(a_t \mid s_t) \qquad \text{(Markov policy)}$$

In general, policies map histories to actions. In certain cases, however, there are optimal policies that are Markov. This is for example the case with additive utility functions $U : \mathcal{R}^* \to \mathbb{R}$, which map the sequence of all possible rewards to a real number as specified in the following definition.

Definition 6.3.2 (*Additive utility*) The utility function $U : \mathcal{R}^* \to \mathbb{R}$ is defined as

$$U(r_0, r_1, \ldots, r_T) = \sum_{k=0}^{T} \gamma^k r_k,$$

horizon

discount factor

where T is the *horizon*, after which the agent is no longer interested in rewards, and $\gamma \in (0, 1]$ is the *discount factor*, which discounts future rewards. It is convenient to introduce a special notation for the utility starting from time t, i.e., the sum of rewards from that time on:

$$U_t \triangleq \sum_{k=0}^{T-t} \gamma^k r_{t+k}.$$

At any time t, the agent wants to to find a policy π *maximizing* the *expected total future reward* (expected utility)

$$\mathbb{E}^\pi_\mu U_t = \mathbb{E}^\pi_\mu \sum_{k=0}^{T-t} \gamma^k r_{t+k}.$$

This is so far identical to the expected utility framework we have seen so far, with the only difference that now the reward space is a sequence of numerical rewards and that we are acting within a dynamical system with state space $\mathcal{S}$. In fact, it is a good idea to think about the *value* of different states of the system under certain policies in the same way that one thinks about how good different positions are in a board game like chess.

6.3.1 Value Functions

Value functions represent the expected utility of a given state or state-action pair for a specific policy. They are useful as shorthand notation and as the basis for algorithm development. The most basic one is the state value function.

State value function

$$V^{\pi}_{\mu,t}(s) \triangleq \mathbb{E}^{\pi}_{\mu}(U_t \mid s_t = s)$$

The state value function for a particular policy π in an MDP μ can be interpreted as how much utility you should expect if you follow the policy starting from state s at time t.

State-action value function

$$Q^{\pi}_{\mu,t}(s, a) \triangleq \mathbb{E}^{\pi}_{\mu}(U_t \mid s_t = s, a_t = a)$$

The state-action value function for a particular policy π in an MDP μ can be interpreted as how much utility you should expect if you play action a at state s at time t, and then follow the policy π. It is also useful to define the *optimal policy* and *optimal value functions* for a given MDP μ. Using a star to indicate optimal quantities, an *optimal policy* π^* dominates all other policies π everywhere in $\mathcal{S}$, that is,

$$V^{\pi^*}_{\mu,t}(s) \geq V^{\pi}_{\mu,t}(s) \quad \forall \pi, t, s.$$

The *optimal value function* V^* is the value function of an optimal policy π^*, i.e.,

$$V^{*}_{\mu,t}(s) \triangleq V^{\pi^*}_{\mu,t}(s), \quad Q^{*}_{\mu,t}(s) \triangleq Q^{\pi^*}_{\mu,t}(s, a).$$

Finding the optimal policy when μ is known

When the MDP μ is known, the expected utility of any policy can be calculated. As the number of policies is exponential in the number of states, it is however not advisable to determine the optimal policy by calculating the utility of every possible policy. More suitable approaches include iterative (offline) methods. These either try to estimate the optimal value function directly, or iteratively improve a policy until it is optimal. Another type of method tries to find an optimal policy in an online fashion. That is, the optimal actions are estimated only for states which can be visited from the current state. However, all these algorithms share the same main ideas.

6.4 Finite Horizon, Undiscounted Problems

The conceptually simplest type of problems are finite horizon problems where $T < \infty$ and $\gamma = 1$. The first thing we shall try is to evaluate a given policy π for a given MDP, that is, compute $V^\pi_{\mu,t}(s)$ for all states s and $t = 0, 1, \ldots T$. There are a number of algorithms that can be used for that purpose.

6.4.1 *Direct Policy Evaluation*

By definition,

$$V^\pi_{\mu,t}(s) \triangleq \mathbb{E}^\pi_\mu(U_t \mid s_t = s) = \sum_{k=0}^{T-t} \mathbb{E}^\pi_\mu(r_{t+k} \mid s_t = s)$$

$$= \sum_{k=t}^{T} \sum_{s' \in \mathcal{S}} \mathbb{E}^\pi_\mu(r_k \mid s_k = s')\, \mathbb{P}^\mu_\pi(s_k = s' \mid s_t = s), \tag{6.4.1}$$

and one can try to compute the value function by (6.4.1), using that

$$\mathbb{P}^\mu_\pi(s_k = s' \mid s_t = s) = \sum_{s'' \in \mathcal{S}} \mathbb{P}(s_k = s' \mid s_{k-1} = s'', s_t = s)\mathbb{P}(s_{k-1} = s'' \mid s_t = s).$$

However, the computational cost of this *direct policy evaluation* is quite high, as it results in a total of $|\mathcal{S}|^3 T$ operations if the value function is to be computed for all time steps up to T.

6.4.2 *Backwards Induction Policy Evaluation*

Noting that

$$\begin{aligned} V^\pi_{\mu,t}(s) &\triangleq \mathbb{E}^\pi_\mu(U_t \mid s_t = s) \\ &= \mathbb{E}^\pi_\mu(r_t \mid s_t = s) + \mathbb{E}^\pi_\mu(U_{t+1} \mid s_t = s) \\ &= \mathbb{E}^\pi_\mu(r_t \mid s_t = s) + \sum_{s' \in \mathcal{S}} p_\mu(s' \mid s, \pi(s))\, V^\pi_{\mu,t+1}(s') \end{aligned}$$

provides a recursion that can be used for the backwards induction algorithm shown as Algorithm 6.1, that is similar to the backwards induction algorithm we have already seen for sequential sampling and bandit problems. However, here in a first step we are only evaluating a given policy π rather than finding the optimal one.

Algorithm 6.1 Backwards induction policy evaluation

input μ, policy π, horizon T
for $s \in S$ **do**
  Initialize $\hat{V}_T(s) = \mathbb{E}^{\pi}_{\mu}(r_T \mid s_T = s)$.
  for $t = T - 1, \ldots, 0$ **do**

$$\hat{V}_t(s) = \mathbb{E}^{\pi}_{\mu}(r_t \mid s_t = s) + \sum_{s' \in \mathcal{S}} p_{\mu}(s' \mid s, \pi(s))\, \hat{V}_{t+1}(s')$$

  end for
end for

Remark 6.4.1 The backwards induction algorithm gives estimates $\hat{V}_t(s)$ satisfying

$$\hat{V}_t(s) = V^{\pi}_{\mu,t}(s).$$

6.4.3 Backwards Induction Policy Optimization

Backwards induction policy optimization as given as Algorithm 6.2 is the first non-naive algorithm for finding an optimal policy for the sequential problem with horizon T. It is basically identical to the backwards induction algorithms we have seen in Chap. 5 for sequential sampling and the bandit problem.

Algorithm 6.2 Finite-horizon backwards induction

input μ, horizon T
Initialize $\hat{V}_T(s) := \max_a r_{\mu}(s, a)$ for all $s \in \mathcal{S}$.
for $t = T - 1, \ldots, 1$ **do**
  for $s \in \mathcal{S}$ **do**
    $\pi_t(s) := \arg\max_a \left\{ r_{\mu}(s, a) + \sum_{s' \in \mathcal{S}} p_{\mu}(s' \mid s, a)\, \hat{V}_{t+1}(s') \right\}$
    $\hat{V}_t(s) := r_{\mu}(s, \pi_t(s)) + \sum_{s' \in \mathcal{S}} p_{\mu}(s' \mid s, \pi_t(s))\, \hat{V}_{t+1}(s')$
  end for
end for
return $\pi = (\pi_t)_{t=1}^{T}$

Theorem 6.4.1 *For T-horizon problems, backwards induction is optimal, i.e.,*

$$\hat{V}_t(s) = V^{*}_{\mu,t}(s).$$

Proof First we show that $\hat{V}_t(s) \geq V^{*}_{\mu,t}(s)$. For $t = T$ we evidently have $\hat{V}_T(s) = \max_a r(s, a) = V^{*}_{\mu,T}(s)$. We proceed by induction assuming that $\hat{V}_{t+1}(s) \geq V^{*}_{\mu,t+1}(s)$ holds. Then for any policy π'

$$\begin{aligned}
\hat{V}_t(s) &= \max_a \left\{ r(s, a) + \sum_{s' \in \mathcal{S}} p_\mu(s' \mid s, a)\, \hat{V}_{t+1}(s') \right\} \\
&\geq \max_a \left\{ r(s, a) + \sum_{s' \in \mathcal{S}} p_\mu(s' \mid s, a)\, V^*_{\mu,t+1}(s') \right\} && \text{(by induction assumption)} \\
&\geq \max_a \left\{ r(s, a) + \sum_{s' \in \mathcal{S}} p_\mu(s' \mid s, a)\, V^{\pi'}_{\mu,t+1}(s') \right\} && \text{(by optimality)} \\
&\geq V^{\pi'}_{\mu,t}(s).
\end{aligned}$$

Choosing $\pi' = \pi^*$ to be the optimal policy this completes the induction proof. Finally, we note that for the policy π returned by backwards induction we have

$$V^*_{\mu,t}(s) \geq V^\pi_{\mu,t}(s) = \hat{V}_t(s) \geq V^*_{\mu,t}(s).$$ □

6.5 Infinite-Horizon

When problems have no fixed horizon, they usually can be modelled as infinite horizon problems, sometimes with help of a *terminal state*, whose visit terminates the problem, or discounted rewards, which indicate that we care less about rewards further in the future. When reward discounting is exponential, these problems can be seen as undiscounted problems with random and geometrically distributed horizon. For problems with no discounting and no terminal states there are some complications in the definition of optimal policy. However, we defer discussion of such problems to Chap. 10.

6.5.1 Examples

We begin with some examples, which will help elucidate the concept of terminal states and infinite horizon.

6.5.1.1 Shortest-Path Problems

First we consider shortest path problems, where the aim is to find the shortest path to a particular goal. Although the process terminates when the goal is reached, not all policies may be able to reach the goal, and so the process may never terminate. We shall consider two types of shortest path problems, deterministic and stochastic. Although conceptually different, both problems have essentially the same complexity.

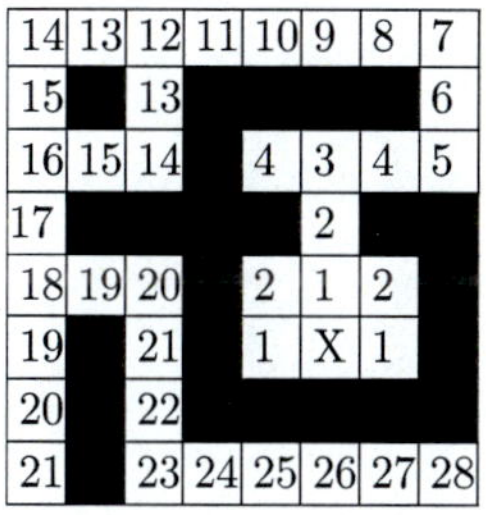

Properties

- $\gamma = 1$, $T \to \infty$.
- $r_t = -1$ unless $s_t = X$, in which case $r_t = 0$.
- $\mathbb{P}_\mu(s_{t+1} = X | s_t = X) = 1$.
- $\mathcal{A} = \{\text{North, South, East, West}\}$
- Transitions are deterministic and walls block.

Fig. 6.4 Deterministic shortest path example

Consider an agent moving in a maze, aiming to get to some terminal goal state X, as for example seen in Fig. 6.4. That is, when reaching this state, the agent stays in X for all further time steps and receives a reward of 0. In general, the agent can move deterministically in the four cardinal directions, and receives a negative reward at each time step. Consequently, the optimal policy is to move to X as quickly as possible.

Solving the shortest path problem with deterministic transitions can be done simply by recursively defining the distance of states to X. Thus, first the distance of X to X is set to 0. Then for states s with distance d to X and with a neighbor state s' with no assigned distance yet, one assigns s' the distance $d + 1$ to X. This is illustrated in Fig. 6.4, where for all reachable states the distance to X is indicated. The respective optimal policy at each step simply moves to a neighbor state with the smallest distance to X. Its reward starting in any state s is simply the negative distance from s to X.

Stochastic shortest path problem with a pit. Now let us assume the shortest path problem with stochastic dynamics. That is, at each time-step there is a small probability ω that we move to a random direction. To make this more interesting, we can add a pit O, that is a terminal state giving a one-time negative reward of -100 (and 0 reward for all further steps) as seen in Fig. 6.5.

Figure 6.6 show that randomness changes the solution significantly in this environment. When ω is relatively small, it is worthwhile (in expectation) for the agent to pass past the pit, even though there is a risk of falling in and getting a reward of -100. In the example given, even starting from the third row in the first column, the agent prefers taking the short-cut. If ω is sufficiently high, the optimal policy avoids however approaching the pit. Still, the agent prefers jumping in the pit (getting a large one-time negative reward) to staying at the save left bottom of the maze forever (with a reward of -1 at each step).

Properties

- $\gamma = 1$, $T \to \infty$.
- $r_t = -1$, but $r_t = 0$ at X and -100 (once, reward 0 afterwards) at O.
- $\mathbb{P}_\mu(s_{t+1} = X | s_t = X) = 1$.
- $\mathbb{P}_\mu(s_{t+1} = O | s_t = O) = 1$.
- $\mathcal{A} = \{\text{North}, \text{South}, \text{East}, \text{West}\}$
- Moves to a random direction with probability ω. Walls block.

Fig. 6.5 Stochastic shortest path example

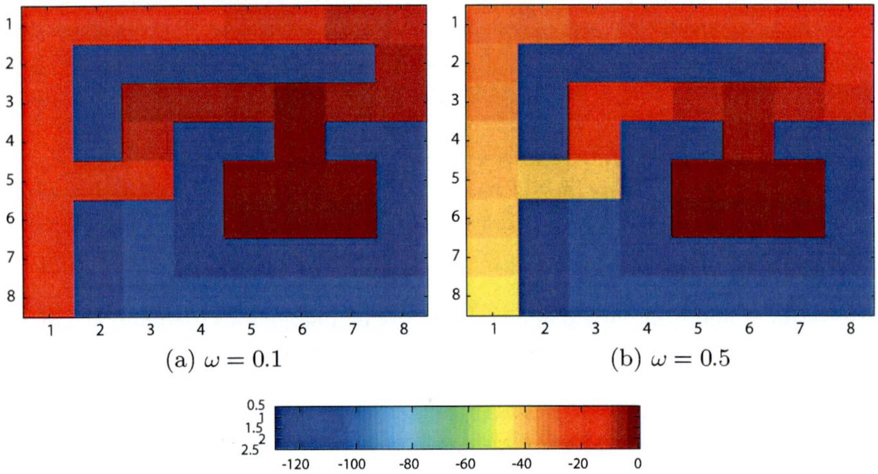

(a) $\omega = 0.1$ (b) $\omega = 0.5$

(c) value

Fig. 6.6 Pit maze solutions for two values of ω

6.5.2 Markov Chain Theory for Discounted Problems

Many problems have no natural terminal state, but continue *ad infinitum*. A popular model to guarantee that the utility is still bounded is to exponentially discount future rewards. This also has some economical interpretation. On the one hand discounting takes into account the effects of inflation, on the other hand money available now may be more useful than money one obtains in the future. Both these effects diminish the value of money over time. In this section we consider some basics of MDPs with infinite horizon and discounted rewards, when the utility is given by

$$U_t = \lim_{T\to\infty} \sum_{k=t}^{T} \gamma^k r_k, \qquad \gamma \in (0, 1).$$

For simplicity, in the following we assume that rewards only depend on the current state instead of both state and action. It can easily be verified that the results presented below can be adapted to the latter case. Henceforth, we will also often drop the dependence on μ in the notation, if the considered MDP μ is clear from the context. As we assume finite state and action spaces $\mathcal{S}$, $\mathcal{A}$ as well as a time-invariant transition kernel we may use the following simplified vector notation:

- $\boldsymbol{r} = (r(s))_{s\in\mathcal{S}}$ is the reward vector in $\mathbb{R}^{|\mathcal{S}|}$.
- We will use $\boldsymbol{P}_\pi$ to denote the transition matrix in $\mathbb{R}^{|\mathcal{S}|\times|\mathcal{S}|}$ for policy π, i.e.,

$$\boldsymbol{P}_\pi(s, s') = \sum_a p(s' \mid s, a)\mathbb{P}^\pi(a \mid s).$$

- $\boldsymbol{v}^\pi = (\mathbb{E}^\pi(U_t \mid s_t = s))_{s\in\mathcal{S}}$ is a vector in $\mathbb{R}^{|\mathcal{S}|}$ representing the value of policy π.

Definition 6.5.1 A policy π is stationary if $\pi(a_t = a \mid s_t = s) = \pi(a_{t'} = a \mid s_{t'} = s)$ for all time steps t, t', states s and actions a.

For infinite-horizon discounted MDPs, considering stationary policies are sufficient in the sense that there is always a stationary policy maximizing utility. This can be proven by induction, using arguments similar to other proofs given here. For a detailed proof see Sect. 5.5 of [3]. The remainder of this chapter collects material from [3]. We now present a set of important results that show how to express MDP quantities like value vectors using linear algebra.

Remark 6.5.1 It holds that

$$\boldsymbol{v}^\pi = \sum_{t=0}^{\infty} \gamma^t \boldsymbol{P}_\pi^t \boldsymbol{r}.$$

Proof Let r_t be the random reward at step t when starting in state s and following policy π. Then

$$
\begin{aligned}
v^\pi(s) &= \mathbb{E}^\pi\left(\sum_{t=0}^{\infty} \gamma^t r_t \,\middle|\, s_0 = s\right) \\
&= \sum_{t=0}^{\infty} \gamma^t \,\mathbb{E}^\pi(r_t \mid s_0 = s) \\
&= \sum_{t=0}^{\infty} \gamma^t \sum_{s' \in \mathcal{S}} \mathbb{P}^\pi(s_t = s' \mid s_0 = s)\ \mathbb{E}(r_t \mid s_t = s') \\
&= \sum_{t=0}^{\infty} \gamma^t \boldsymbol{P}_\pi^t \boldsymbol{r},
\end{aligned}
$$

as the entries of $\boldsymbol{P}_\pi^t$ are precisely the t-step transition probabilities, and for any distribution vector $\boldsymbol{p}$ over $\mathcal{S}$, we have $\mathbb{E}_{\boldsymbol{p}}\, r_t = \boldsymbol{p}^\top \boldsymbol{r}_t$. □

One can show that the expected discounted total reward of a policy is equal to the expected undiscounted total reward with a geometrically distributed horizon (see Exercise 6.1). Accordingly, an MDP with discounted rewards is equivalent to one where there is no discounting but a stopping probability $(1 - \gamma)$ at every step.

The value of a particular policy can be expressed as a linear equation. This is an important result, that has led to a number of successful algorithms that employ linear algebra.

Theorem 6.5.1 *For any stationary policy π, $\boldsymbol{v}^\pi$ is the unique solution of*

$$\boldsymbol{v} = \boldsymbol{r} + \gamma \boldsymbol{P}_\pi \boldsymbol{v}. \tag{6.5.1}$$

In addition, the solution is

$$\boldsymbol{v}^\pi = (\boldsymbol{I} - \gamma \boldsymbol{P}_\pi)^{-1} \boldsymbol{r},$$

where $\boldsymbol{I}$ is the identity matrix.

For the proof of Theorem 6.5.1 we make a few preparations first. For a matrix $\boldsymbol{A} \in \mathbb{R}^{n \times n}$ with entries a_{ij} we define the *norm*

$$\|\boldsymbol{A}\| \triangleq \max_i \sum_j |a_{i,j}|$$

spectral radius

and the *spectral radius*

$$\sigma(\boldsymbol{A}) \triangleq \lim_{t \to \infty} \|\boldsymbol{A}^t\|^{1/t}.$$

It holds that $\|\boldsymbol{A}\| \geq \sigma(\boldsymbol{A})$, and if $\boldsymbol{A}$ is a probability matrix then $\|\boldsymbol{A}\| = \sigma(\boldsymbol{A}) = 1$. Further, the following theorem holds.

Theorem 6.5.2 *If $\sigma(\boldsymbol{A}) < 1$, then $(\boldsymbol{I} - \boldsymbol{A})^{-1}$ exists and is given by*

$$(\boldsymbol{I} - \boldsymbol{A})^{-1} = \lim_{T \to \infty} \sum_{t=0}^{T} \boldsymbol{A}^t. \tag{6.5.2}$$

Proof of Theorem 6.5.1 *First note that from (6.5.1) one obtains $\boldsymbol{r} = (\boldsymbol{I} - \gamma \boldsymbol{P}_\pi)\boldsymbol{v}$. Since $\|\gamma \boldsymbol{P}_\pi\| = \gamma \cdot \|\boldsymbol{P}_\pi\| = \gamma < 1$, the inverse*

$$(\boldsymbol{I} - \gamma \boldsymbol{P}_\pi)^{-1} = \lim_{T \to \infty} \sum_{t=0}^{T} (\gamma \boldsymbol{P}_\pi)^t$$

exists by Theorem 6.5.2. It follows that

$$\boldsymbol{v} = (\boldsymbol{I} - \gamma \boldsymbol{P}_\pi)^{-1} \boldsymbol{r} = \sum_{t=0}^{\infty} \gamma^t \boldsymbol{P}_\pi^t \boldsymbol{r} = \boldsymbol{v}^\pi,$$

where the last step is by Remark 6.5.1. □

It is important to note that the entries of the matrix $\boldsymbol{X} = (\boldsymbol{I} - \gamma \boldsymbol{P}_\pi)^{-1}$ are the expected number of discounted cumulative visits to each state s, starting from state s' and following policy π. That is,

$$x(s, s') = \mathbb{E}_\mu^\pi \left\{ \sum_{t=0}^{\infty} \gamma^t \mathbb{P}_\mu^\pi (s_t = s' \mid s_t = s) \right\}.$$

This interpretation is quite useful, as many algorithms rely on an estimation of $\boldsymbol{X}$ for approximating value functions.

6.5.3 Optimality Equations

Let us now look at the backwards induction algorithms in terms of operators. We first introduce the one-step backwards induction operator for a fixed policy and the Bellman operator, which for the optimal policy coincides with the former operator. In the following, we denote by $\mathcal{V}$ the space of value functions, which is a Banach space (i.e., a complete, normed vector space) equipped with the norm

$$\|\boldsymbol{v}\| = \max \left\{ |\boldsymbol{v}(s)| \middle| s \in \mathcal{S} \right\}.$$

Definition 6.5.2 (*Policy and Bellman operator*) Let $\boldsymbol{v} \in \mathcal{V}$. Then the linear operator of a policy π is given by

$$\mathscr{L}_\pi \boldsymbol{v} \triangleq \boldsymbol{r} + \gamma \boldsymbol{P}_\pi \boldsymbol{v}.$$

Further, the (non-linear) Bellman operator is defined as

$$\mathscr{L}\boldsymbol{v} \triangleq \max_{\pi} \{\boldsymbol{r} + \gamma \boldsymbol{P}_{\pi} \boldsymbol{v}\}.$$

We now show that the Bellman operator satisfies the following monotonicity properties with respect to an arbitrary value function $\boldsymbol{v}$. Further, if a value function $\boldsymbol{v}$ is optimal, then it satisfies the *Bellman optimality equation*

Bellman optimality equation

$$\boldsymbol{v} = \mathscr{L}\boldsymbol{v}.$$

In the following, we use the notation $\boldsymbol{v} \geq \boldsymbol{v}'$ in short for $\boldsymbol{v}(s) \geq \boldsymbol{v}'(s)$ for all s.

Theorem 6.5.3 *Let $\boldsymbol{v}^* \triangleq \max_{\pi} \boldsymbol{v}^{\pi}$. Then for any $\boldsymbol{v} \in \mathcal{V}$:*

(1) If $\boldsymbol{v} \geq \mathscr{L}\boldsymbol{v}$, then $\boldsymbol{v} \geq \boldsymbol{v}^*$.
(2) If $\boldsymbol{v} \leq \mathscr{L}\boldsymbol{v}$, then $\boldsymbol{v} \leq \boldsymbol{v}^*$.
(3) If $\boldsymbol{v} = \mathscr{L}\boldsymbol{v}$, then $\boldsymbol{v} = \max_{\pi} \boldsymbol{v}^{\pi}$.

Proof We first prove (1). A simple proof by induction over n shows that for any π and any n

$$\boldsymbol{v} \geq \boldsymbol{r} + \gamma \boldsymbol{P}_{\pi} \boldsymbol{v} \geq \sum_{t=0}^{n-1} \gamma^t \boldsymbol{P}_{\pi}^t \boldsymbol{r} + \gamma^n \boldsymbol{P}_{\pi}^n \boldsymbol{v}.$$

Since $\boldsymbol{v}^{\pi} = \sum_{t=0}^{\infty} \gamma^t \boldsymbol{P}_{\pi}^t \boldsymbol{r}$, it follows that

$$\boldsymbol{v} - \boldsymbol{v}^{\pi} \geq \gamma^n \boldsymbol{P}_{\pi}^n \boldsymbol{v} - \sum_{t=n}^{\infty} \gamma^t \boldsymbol{P}_{\pi}^t \boldsymbol{r}.$$

Let $\boldsymbol{e}$ be the vector with $\boldsymbol{e}(s) = 1$ for all s. Then, as rewards are assumed to be in $[0, 1]$,

$$\sum_{k=n}^{\infty} \gamma^k \boldsymbol{P}_{\pi}^k \boldsymbol{r} \leq \frac{\gamma^n \boldsymbol{e}}{1 - \gamma},$$

which for $n \to \infty$ approaches $\boldsymbol{0}$. It follows that for any π

$$\boldsymbol{v} \geq \boldsymbol{v}^{\pi},$$

and hence also $\boldsymbol{v} \geq \boldsymbol{v}^*$, which completes the proof of (1). A similar argument shows (2), which together with (1) then implies (3). □

A similar theorem can also be proven for the repeated application of the Bellman operator.

Theorem 6.5.4 *1. Let $\boldsymbol{v}, \boldsymbol{v}' \in \mathcal{V}$ with $\boldsymbol{v}' \geq \boldsymbol{v}$. Then $\mathscr{L}\boldsymbol{v}' \geq \mathscr{L}\boldsymbol{v}$.*
2. Consider a sequence $(\boldsymbol{v}_n)_{n\geq 0}$ of values functions with arbitrary $\boldsymbol{v}_0 \in \mathcal{V}$ and $\boldsymbol{v}_{n+1} = \mathscr{L}\boldsymbol{v}_n$ for $n \geq 0$. If there is an N with $\mathscr{L}\boldsymbol{v}_N \leq \boldsymbol{v}_N$, then $\mathscr{L}\boldsymbol{v}_{N+k} \leq \boldsymbol{v}_{N+k}$ for all $k \geq 0$ and similarly for $\geq$.

Proof Let $\pi \in \arg\max_\pi \boldsymbol{r} + \gamma \boldsymbol{P}_\pi \boldsymbol{v}$. Then

$$\mathscr{L}\boldsymbol{v} = \boldsymbol{r} + \gamma \boldsymbol{P}_\pi \boldsymbol{v} \leq \boldsymbol{r} + \gamma \boldsymbol{P}_\pi \boldsymbol{v}' \leq \max_{\pi'} \boldsymbol{r} + \gamma \boldsymbol{P}_{\pi'} \boldsymbol{v}' = \mathscr{L}\boldsymbol{v}',$$

where the first inequality is due to the fact that $\boldsymbol{P}\boldsymbol{v} \geq \boldsymbol{P}\boldsymbol{v}'$ for any transition matrix $\boldsymbol{P}$. For the second part, note that we have $\mathscr{L}\boldsymbol{v}_N \leq \boldsymbol{v}_N$ by assumption and hence $\mathscr{L}^k \mathscr{L}\boldsymbol{v}_N \leq \mathscr{L}^k \boldsymbol{v}_N$ by the first part of the theorem. It follows that

$$\mathscr{L}\boldsymbol{v}_{N+k} = \boldsymbol{v}_{N+k+1} = \mathscr{L}^k \mathscr{L}\boldsymbol{v}_N \leq \mathscr{L}^k \boldsymbol{v}_N = \boldsymbol{v}_{N+k}. \qquad \square$$

Thus, if one starts with an initial value $\boldsymbol{v}_0 \leq \boldsymbol{v}'$ for all $\boldsymbol{v}' \in \mathcal{V}$, then repeated application of the Bellman operator (known as *value iteration* as introduced in the following section) converges monotonically to $\boldsymbol{v}^*$. For example, if rewards are ≥ 0, one may set $\boldsymbol{v}_0 = \boldsymbol{0}$ and $\boldsymbol{v}_n = \mathscr{L}^n \boldsymbol{v}_0$ is always a lower bound on the optimal value function.

We eventually want to show that repeated application of the Bellman operator always converges to the optimal value, independent of the initial value $\boldsymbol{v}_0$. As a preparation, we need the following definition and the subsequent theorem.

Definition 6.5.3 For a Banach space $\mathcal{X}$ (i.e., a complete, normed linear space) we say that $T : \mathcal{X} \to \mathcal{X}$ is a *contraction mapping*, if there is $\gamma \in [0, 1)$ such that $\|T\boldsymbol{u} - T\boldsymbol{v}\| \leq \gamma \|\boldsymbol{u} - \boldsymbol{v}\|$ for all $\boldsymbol{u}, \boldsymbol{v} \in \mathcal{X}$.

Theorem 6.5.5 (Banach fixed-point theorem) *Given a Banach space $\mathcal{X}$ and a contraction mapping $T : \mathcal{X} \to \mathcal{X}$, it holds that:*

1. There is a unique $\boldsymbol{u}^* \in \mathcal{X}$ with $T\boldsymbol{u}^* = \boldsymbol{u}^*$.
2. For any $\boldsymbol{u}^0 \in \mathcal{X}$ the sequence $(\boldsymbol{u}_n)_{n\geq 0}$ defined by $\boldsymbol{u}_{n+1} = T\boldsymbol{u}_n = T^{n+1}\boldsymbol{u}_0$ converges to $\boldsymbol{u}^*$.

Proof First note that for any $m \geq 1$

$$\begin{aligned} \|\boldsymbol{u}_{n+m} - \boldsymbol{u}_n\| &\leq \sum_{k=0}^{m-1} \|\boldsymbol{u}_{n+k+1} - \boldsymbol{u}_{n+k}\| = \sum_{k=0}^{m-1} \|T^{n+k}\boldsymbol{u}_1 - T^{n+k}\boldsymbol{u}_0\| \\ &\leq \sum_{k=0}^{m-1} \gamma^{n+k} \|\boldsymbol{u}_1 - \boldsymbol{u}_0\| = \frac{\gamma^n(1-\gamma^m)}{1-\gamma} \|\boldsymbol{u}_1 - \boldsymbol{u}_0\|. \end{aligned}$$

Therefore, for each $\epsilon > 0$, there is n such that $\|\boldsymbol{u}^{n+m} - \boldsymbol{u}^n\| \leq \epsilon$. Since $\mathcal{X}$ is a Banach space, the sequence has a limit $\boldsymbol{u}^* \in \mathcal{X}$.

Next, we show that this $\boldsymbol{u}^*$ is indeed a fixed point of T. Indeed, we have

$$\|T\boldsymbol{u}^* - \boldsymbol{u}^*\| \le \|T\boldsymbol{u}^* - \boldsymbol{u}_n\| + \|\boldsymbol{u}_n - \boldsymbol{u}^*\|. \tag{6.5.3}$$

As $\|T\boldsymbol{u}^* - \boldsymbol{u}_n\| = \|T\boldsymbol{u}^* - T\boldsymbol{u}_{n-1}\| \le \gamma\|\boldsymbol{u}^* - \boldsymbol{u}_{n-1}\|$, both terms on the right hand side of (6.5.3) approach 0 for $n \to \infty$, whence we have $\|T\boldsymbol{u}^* - \boldsymbol{u}^*\| = 0$ and $T\boldsymbol{u}^* = \boldsymbol{u}^*$.

We conclude by showing uniqueness. If $\boldsymbol{u}'$, $\boldsymbol{u}''$ are both fixed points then by the contraction property

$$\|\boldsymbol{u}' - \boldsymbol{u}''\| = \|T\boldsymbol{u}' - T\boldsymbol{u}''\| \le \gamma\|\boldsymbol{u}' - \boldsymbol{u}''\|,$$

whence it follows that $\|\boldsymbol{u}' - \boldsymbol{u}''\| = 0$ and $\boldsymbol{u}' = \boldsymbol{u}''$. □

Theorem 6.5.6 *For $\gamma \in [0, 1)$ the Bellman operator $\mathscr{L}$ is a contraction mapping in $\mathcal{V}$.*

Proof Let $\boldsymbol{v}, \boldsymbol{v}' \in \mathcal{V}$. Consider $s \in \mathcal{S}$ such that $\mathscr{L}\boldsymbol{v}(s) \ge \mathscr{L}\boldsymbol{v}'(s)$, and let

$$a_s^* \in \arg\max_{a \in \mathcal{A}} \left\{ r(s) + \sum_{s' \in \mathcal{S}} \gamma\, p(s' \mid s, a)\, \boldsymbol{v}(s') \right\}.$$

Then, as a_s^* is optimal for $\boldsymbol{v}(s)$, but not necessarily for $\boldsymbol{v}'(s)$, we have

$$\begin{aligned}
0 \le \mathscr{L}\boldsymbol{v}(s) - \mathscr{L}\boldsymbol{v}'(s) &\le \sum_{s' \in S} \gamma\, p(s' \mid s, a_s^*)\, \boldsymbol{v}(s') - \sum_{s' \in S} \gamma\, p(s' \mid s, a_s^*)\, \boldsymbol{v}'(s') \\
&= \gamma \sum_{s' \in S} p(s' \mid s, a_s^*)\, [\boldsymbol{v}(s') - \boldsymbol{v}'(s')] \\
&\le \gamma \sum_{s' \in S} p(s' \mid s, a_s^*)\, \|\boldsymbol{v} - \boldsymbol{v}'\| = \gamma\|\boldsymbol{v} - \boldsymbol{v}'\|.
\end{aligned}$$

Repeating the argument for s such that $\mathscr{L}\boldsymbol{v}(s) \le \mathscr{L}\boldsymbol{v}'(s)$, we obtain

$$|\mathscr{L}\boldsymbol{v}(s) - \mathscr{L}\boldsymbol{v}'(s)| \le \gamma\|\boldsymbol{v} - \boldsymbol{v}'\|.$$

It follows that $\|\mathscr{L}\boldsymbol{v} - \mathscr{L}\boldsymbol{v}'\| = \max_s |\mathscr{L}\boldsymbol{v}(s) - \mathscr{L}\boldsymbol{v}'(s)| \le \gamma\|\boldsymbol{v} - \boldsymbol{v}'\|$. □

We note that is easy to adapt this proof to show that $\mathscr{L}_\pi$ is a contraction, too.

Theorem 6.5.7 *1. There is a unique $\boldsymbol{v}^* \in \mathcal{V}$ such that $\mathscr{L}\boldsymbol{v}^* = \boldsymbol{v}^*$ and $\boldsymbol{v}^* = \max_\pi \boldsymbol{v}^\pi$.*
2. For any stationary policy π, there is a unique $\boldsymbol{v} \in \mathcal{V}$ such that $\mathscr{L}_\pi\boldsymbol{v} = \boldsymbol{v}$ and $\boldsymbol{v} = \boldsymbol{v}^\pi$.

Proof As the Bellman operator $\mathscr{L}$ is a contraction by Theorem 6.5.6, application of Theorem 6.5.5 shows that there is a unique $\boldsymbol{v}^* \in \mathcal{V}$ such that $\mathscr{L}\boldsymbol{v}^* = \boldsymbol{v}^*$. This is also the optimal value function due to Theorem 6.5.3. The second part of the theorem follows from the first part when considering only a single policy π (which then is optimal). □

6.5.4 MDP Algorithms for Infinite Horizon and Discounted Rewards

We now take a look at the basic algorithms for obtaining an optimal policy when the Markov decision process is known. *Value iteration* is a simple extension of the backwards induction algorithm to the infinite horizon case. Alternatively, *policy iteration* evaluates and improves a sequence of Markov policies. We also discuss two variants of these methods, namely *modified policy iteration*, which is somewhere in between value and policy iteration, and *temporal-difference policy iteration*, which is related to classical reinforcement learning algorithms such as Sarsa and Q-Learning. Another basic technique is *linear programming*, which is useful in theoretical analyses as well as in some special practical cases. While for the sake of simplicity, we stick to our assumption that rewards depend only on the state, the algorithms described below can easily extended to the case when the reward also depends on the action.

6.5.4.1 Value Iteration

Value iteration corresponds to repeated application of the Bellman operator introduced in the previous section.

Algorithm 6.3 Value iteration

```
input μ
Initialization: Choose some v_0 ∈ V.
for k = 1, 2, ..., n do
  for s ∈ S do
    v_k(s) = max_a {r(s) + γ Σ_{s'∈S} p(s' | s, a) v_{k-1}(s')}
  end for
end for
return v_n and π_n = arg max_π {r + γ P_π v_n}
```

Value iteration is a direct extension of the backwards induction algorithm for infinite horizon and finite state space $\mathcal{S}$. Since we know that stationary policies are optimal, there is no need to maintain the values and actions for all time steps. At each step, it is sufficient to merely keep the previous value $\boldsymbol{v}_{n-1}$. The following theorem guarantees that value iteration converges to the optimal value and provides bounds for the error of each single estimate $\boldsymbol{v}_n$.

Theorem 6.5.8 *Value iteration satisfies:*

1. $\lim_{n\to\infty} \|\boldsymbol{v}_n - \boldsymbol{v}^*\| = 0$.
2. *For each* $\epsilon > 0$ *there exists* $N_\epsilon < \infty$ *such that for all* $n \geq N_\epsilon$*:*
   (a) $\|\boldsymbol{v}_{n+1} - \boldsymbol{v}_n\| \leq \epsilon(1-\gamma)/2\gamma$.
   (b) $\|\boldsymbol{v}_n - \boldsymbol{v}^*\| < \epsilon/2$.
   (c) *The policy* π_n *is* ϵ*-optimal, i.e.,* $\boldsymbol{v}^{\pi_n}(s) \geq \boldsymbol{v}^*(s) - \epsilon$ *for all states* s.

Proof Statements 1 and 2(a) follow from Theorems 6.5.5 and 6.5.6 from the previous section. Now note that

$$\|\boldsymbol{v}^{\pi_n} - \boldsymbol{v}^*\| \leq \|\boldsymbol{v}^{\pi_n} - \boldsymbol{v}_n\| + \|\boldsymbol{v}_n - \boldsymbol{v}^*\|.$$

For the first term we have by Theorems 6.5.1 and 6.5.6

$$\begin{aligned}
\|\boldsymbol{v}^{\pi_n} - \boldsymbol{v}_n\| &= \left\|\mathscr{L}_{\pi_n}\boldsymbol{v}^{\pi_n} - \boldsymbol{v}_{n+1}\right\| \\
&\leq \left\|\mathscr{L}_{\pi_n}\boldsymbol{v}^{\pi_n} - \mathscr{L}\boldsymbol{v}_{n+1}\right\| + \|\mathscr{L}\boldsymbol{v}_{n+1} - \boldsymbol{v}_{n+1}\| \\
&= \left\|\mathscr{L}_{\pi_n}\boldsymbol{v}^{\pi_n} - \mathscr{L}_{\pi_n}\boldsymbol{v}_{n+1}\right\| + \|\mathscr{L}\boldsymbol{v}_{n+1} - \mathscr{L}\boldsymbol{v}_n\| \\
&\leq \gamma\,\|\boldsymbol{v}^{\pi_n} - \boldsymbol{v}_{n+1}\| + \gamma\,\|\boldsymbol{v}_{n+1} - \boldsymbol{v}_n\|.
\end{aligned}$$

A similar argument gives a respective bound for the second term $\|\boldsymbol{v}_n - \boldsymbol{v}^*\|$. Then, rearranging we obtain

$$\|\boldsymbol{v}^{\pi_n} - \boldsymbol{v}_{n+1}\| \leq \frac{\gamma}{1-\gamma}\|\boldsymbol{v}_{n+1} - \boldsymbol{v}_n\|, \qquad \|\boldsymbol{v}_n - \boldsymbol{v}^*\| \leq \frac{\gamma}{1-\gamma}\|\boldsymbol{v}_{n+1} - \boldsymbol{v}_n\|,$$

and 2(b) as well as 2(c) follow from 2(a). □

Theorem 6.5.8 also bounds the error when stopping value iteration when the change in the estimated value function from one iteration to the next falls below a set small threshold. As we have already discussed in context of Theorem 6.5.4 in the previous section, initializing $\boldsymbol{v}_0 = \boldsymbol{0}$ guarantees that value iteration converges monotonically. The following theorem shows that value iteration converges converges with an error of $O(\gamma^n)$ in this case.

Theorem 6.5.9 *Let* $\boldsymbol{v}_0 = \boldsymbol{0}$ *and assume that rewards are bounded in* $[0, 1]$*. Then*

$$\|\boldsymbol{v}_n - \boldsymbol{v}^*\| \leq \frac{\gamma^n}{1-\gamma}, \quad \textit{and} \quad \|\boldsymbol{v}^{\pi_n} - \boldsymbol{v}^*\| \leq \frac{2\gamma^n}{1-\gamma}.$$

Proof For the first part, note that

$$\|\boldsymbol{v}_0 - \boldsymbol{v}^*\| = \|\boldsymbol{v}^*\| \leq \frac{1}{1-\gamma} = \frac{\gamma^0}{1-\gamma}.$$

Proceeding by induction over n proves the first claim, as by the contraction property of Theorem 6.5.6 we have

$$\|\boldsymbol{v}_{n+1} - \boldsymbol{v}^*\| = \|\mathscr{L}\boldsymbol{v}_n - \mathscr{L}\boldsymbol{v}^*\| \leq \gamma\|\boldsymbol{v}_n - \boldsymbol{v}^*\|.$$

The second part can be shown similarly to 2(c) of Theorem 6.5.8. □

Although value iteration converges exponentially fast, the convergence depends on the discount factor γ. When γ is very close to one, convergence can be extremely slow. In fact, [4] showed that the number of iterations are of the order $1/(1-\gamma)$ for bounded accuracy of the input data. Omitting logarithmic factors the overall complexity is $\tilde{O}(|\mathcal{S}|^2|\mathcal{A}|L(1-\gamma)^{-1})$, where L is the total number of bits used to represent the input.[1]

6.5.4.2 Policy Iteration

Unlike value iteration, *policy iteration* attempts to iteratively improve a given policy, rather than a value function. Starting with an arbitrary policy π_0, at each iteration it first computes the value of the current policy. In finite MDPs this *policy evaluation* step can be performed with either linear algebra or backwards induction. In a second step called *policy improvement* the policy is updated by choosing the policy that is greedy with respect to the value function computed in the evaluation step.

Algorithm 6.4 Policy iteration

input μ
Initialization: Choose some policy π_0. Set $n = 0$.
repeat
  $\boldsymbol{v}_n = \boldsymbol{v}^{\pi_n}$ `// policy evaluation`
  $\pi_{n+1} = \arg\max_\pi \{\boldsymbol{r} + \gamma \boldsymbol{P}_\pi \boldsymbol{v}_n\}$ `// policy improvement`
  n = n + 1
until $\pi_{n+1} = \pi_n$
return $\pi_{n+1}, \boldsymbol{v}_n$

The following theorem shows that the policies generated by policy iteration are monotonically improving.

Theorem 6.5.10 *For value vectors $\boldsymbol{v}_n, \boldsymbol{v}_{n+1}$ generated by policy iteration it holds that $\boldsymbol{v}_n \leq \boldsymbol{v}_{n+1}$.*

Proof From the policy improvement step

$$\boldsymbol{r} + \gamma \boldsymbol{P}_{\pi_{n+1}} \boldsymbol{v}_n \geq \boldsymbol{r} + \gamma \boldsymbol{P}_{\pi_n} \boldsymbol{v}_n = \boldsymbol{v}_n$$

[1] Thus, the result is *weakly* polynomial complexity, due to the dependence on the input size description.

where the equality is due to the policy evaluation step for π_n. Rearranging, we get $\boldsymbol{r} \geq (\boldsymbol{I} - \gamma \boldsymbol{P}_{\pi_{n+1}})\, \boldsymbol{v}_n$ and hence

$$(\boldsymbol{I} - \gamma \boldsymbol{P}_{\pi_{n+1}})^{-1} \boldsymbol{r} \geq \boldsymbol{v}_n.$$

By Theorem 6.5.1 and the policy evaluation step for π_{n+1} the left hand side equals $\boldsymbol{v}_{n+1}$, which completes the proof.

Theorem 6.5.10 can be used to show that policy iteration terminates after a finite number of steps.

Corollary 6.5.1 *Policy iteration terminates after a finite number of iterations and returns an optimal policy.*

Proof There is only a finite number of policies, and since policies in policy iteration are monotonically improving, the algorithm must stop after finitely many iterations. Finally, the last iteration satisfies

$$\boldsymbol{v}_n = \max_{\pi} \{\boldsymbol{r} + \gamma \boldsymbol{P}_\pi \boldsymbol{v}_n\},$$

that is, $\boldsymbol{v}_n$ solves the optimality equation and the claim follows by Theorem 6.5.3. □

As even in finite MDPs, there are $|\mathcal{A}|^{|\mathcal{S}|}$ policies, Corollary 6.5.1 only guarantees exponential-time convergence in the number of states. However, the complexity of policy iteration can be shown to be actually strongly polynomial [5] with the number of required iterations being $\tilde{O}(|\mathcal{S}|^2 |\mathcal{A}| (1-\gamma)^{-1})$, again omitting logarithmic factors.

As the behaviour of policy iteration seems to be quite different from value iteration, one is interested in algorithms that lie between policy iteration and value iteration. We will have a look at two such algorithms in the following two subsections.

6.5.4.3 Modified Policy Iteration

Modified policy iteration tries to speed up policy iteration by performing an m-step update in the policy evaluation step. For $m = 1$, the algorithm is identical to value iteration, while for $m \to \infty$ the algorithm corresponds to policy iteration. Modified policy iteration can perform much better than either pure value iteration or pure policy iteration.

Algorithm 6.5 Modified policy iteration

input μ, parameter m
Initialization: Choose some $\boldsymbol{v}_0 \in \mathcal{V}$.
for $k = 1, 2, \ldots, n$ **do**
  $\pi_k = \arg\max_\pi \{\boldsymbol{r} + \gamma \boldsymbol{P}_\pi \boldsymbol{v}_{k-1}\}$ `// policy improvement`
  $\boldsymbol{v}_k = \mathscr{L}_{\pi_k}^m \boldsymbol{v}_{k-1}$ `// partial policy evaluation`
end for
return $\pi_n, \boldsymbol{v}_n$

6.5.4.4 A Geometric View

It is perhaps interesting to see the problem from a geometric perspective. This also gives rise to the so-called *temporal-difference* algorithms which will be considered below. First, we define the difference operator, which is the difference between a value function vector $\boldsymbol{v}$ and its transformation via the Bellman operator.

Definition 6.5.4 The *difference operator* is defined as

difference operator

$$\mathscr{B}\boldsymbol{v} \triangleq \max_{\pi} \{\boldsymbol{r} + (\gamma \boldsymbol{P}_\pi - \boldsymbol{I})\boldsymbol{v}\} = \mathscr{L}\boldsymbol{v} - \boldsymbol{v}.$$

Accordingly, the Bellman optimality equation can be rewritten as

$$\mathscr{B}\boldsymbol{v} = \boldsymbol{0}.$$

Defining the set of greedy policies with respect to a value vector $\boldsymbol{v} \in \mathcal{V}$ as

$$\Pi_{\boldsymbol{v}} \triangleq \arg\max_{\pi} \{\boldsymbol{r} + (\gamma \boldsymbol{P}_\pi - \boldsymbol{I})\boldsymbol{v}\},$$

we can show the following inequality between two value function vectors.

Theorem 6.5.11 *For any $\boldsymbol{v}, \boldsymbol{v}' \in \mathcal{V}$ and $\pi \in \Pi_{\boldsymbol{v}}$*

$$\mathscr{B}\boldsymbol{v}' \geq \mathscr{B}\boldsymbol{v} + (\gamma \boldsymbol{P}_\pi - \boldsymbol{I})(\boldsymbol{v}' - \boldsymbol{v}).$$

Proof By definition, $\mathscr{B}\boldsymbol{v}' \geq \boldsymbol{r} + (\gamma \boldsymbol{P}_\pi - \boldsymbol{I})\boldsymbol{v}'$, while $\mathscr{B}\boldsymbol{v} = \boldsymbol{r} + (\gamma \boldsymbol{P}_\pi - \boldsymbol{I})\boldsymbol{v}$. Subtracting the latter from the former gives the result. □

The inequality in Theorem 6.5.11 is similar to the convexity of the Bayes-optimal utility (3.3.2). Geometrically, we can see from a look at Fig. 6.7, that applying the Bellman operator on value function always improves it, yet may have a negative effect on the other value function. If the number of policies is finite, then the figure is also a good illustration of the policy iteration algorithm, where each value function improvement results in a new point on the horizontal axis, and the choice of the best improvement (highest line) for that point. In fact, we can write the policy iteration algorithm in terms of the difference operator.

Theorem 6.5.12 *Let $(\boldsymbol{v}_n)_{n \geq 0}$ be the sequence of value vectors obtained from policy iteration. Then for any $\pi \in \Pi_{\boldsymbol{v}_n}$,*

$$\boldsymbol{v}_{n+1} = \boldsymbol{v}_n - (\gamma \boldsymbol{P}_\pi - \boldsymbol{I})^{-1}\mathscr{B}\boldsymbol{v}_n. \tag{6.5.4}$$

Proof By definition, we have for $\pi \in \Pi_{\boldsymbol{v}_n}$

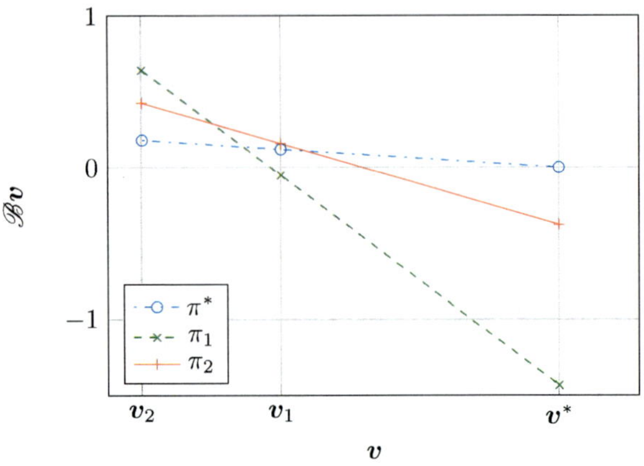

Fig. 6.7 The graph shows the effect of the difference operator on the optimal value function v^* as well as on two arbitrary value functions, v_1, v_2. Each line is the improvement effected by the greedy policy π^*, π_1, π_2 with respect to each value function v^*, v_1, v_2

$$\begin{aligned} v_{n+1} &= (I - \gamma P_\pi)^{-1} r - v_n + v_n \\ &= (I - \gamma P_\pi)^{-1} [r - (I - \gamma P_\pi) v_n] + v_n. \end{aligned}$$

Since $r - (I - \gamma P_\pi) v_n = \mathscr{B} v_n$, the claim follows. □

6.5.4.5 Temporal-Difference Policy Iteration

Similarly to modified policy iteration, *temporal-difference policy iteration* replaces the next-step value with an approximation v_n of the value v^{π_n} of the policy at step n. Informally, this approximation is chosen so as to reduce the discrepancy of the value function over time.

More precisely, at the nth iteration of the algorithm, we use a policy improvement step to obtain the next policy π_{n+1} given the current approximation v_n, i.e.,

$$\mathscr{L}_{\pi_{n+1}} v_n = \mathscr{L} v_n.$$

temporal difference

In order to update the value from v_n to v_{n+1} we rely on the *temporal difference* defined as

$$d_n(s, s') = \big(r(s) + \gamma v_n(s')\big) - v_n(s).$$

The temporal difference error can be seen as the difference in the estimate when we move from state s to state s'. In fact, it is easy to see that if the value function estimate satisfies $v_n = v^{\pi_n}$, then the expected error is zero, as

$$\sum_{s' \in \mathcal{S}} d_n(s, s')\, p(s' \mid s, \pi_n(s)) = \sum_{s' \in \mathcal{S}} \big(r(s) + \gamma v_n(s')\big)\, p(s' \mid s, \pi_n(s)) - v_n(s).$$

Note the similarity to the difference operator in modified policy iteration. The idea of temporal-difference policy iteration is to adjust the current value $\boldsymbol{v}_n$, using the temporal differences mixed over an infinite number of steps, that is, we set

$$\boldsymbol{v}_{n+1} = \boldsymbol{v}_n + \boldsymbol{\tau}_n, \text{ where}$$
$$\boldsymbol{\tau}_n(s) = \sum_{t=0}^{\infty} \mathbb{E}_{\pi_n}\big[(\gamma\lambda)^t d_n(s_t, s_{t+1}) \mid s_0 = s\big].$$

The parameter λ is a simple way to weight the different temporal difference errors. If $\lambda \to 0$, the error $\boldsymbol{\tau}_n$ is dominated by the short-term discrepancies in our value function, while for $\lambda \to 1$ also the terms far in the future matter. In the end, the value function is adjusted in the direction of this error.

Putting all of those steps together, the algorithm looks as follows.

Algorithm 6.6 Temporal-Difference Policy Iteration

input μ, parameter $\lambda \in [0, 1]$
Initialization: Choose some $\boldsymbol{v}_0 \in \mathcal{V}$. Set $n = 0$.
repeat
  $\pi_{n+1} = \arg\max_\pi \{\boldsymbol{r} + \gamma \boldsymbol{P}_\pi \boldsymbol{v}_n\}$ `// policy improvement`
  $\boldsymbol{v}_{n+1} = \boldsymbol{v}_n + \boldsymbol{\tau}_n$ `// temporal difference update`
  $n = n + 1$
until $\pi_{n+1} = \pi_n$
return $\pi_n, \boldsymbol{v}_n$

It can be shown that $\boldsymbol{v}_{n+1}$ is the unique fixed point of the equation

$$\mathscr{D}_n \boldsymbol{v} \triangleq (1-\lambda)\mathscr{L}_{\pi_{n+1}} \boldsymbol{v}_n + \lambda \mathscr{L}_{\pi_{n+1}} \boldsymbol{v}.$$

That is, if we repeatedly apply the above operator to some vector $\boldsymbol{v}$, then we approach a fixed point $\boldsymbol{v}^* = \mathscr{D}_n \boldsymbol{v}^*$. It is interesting to see what happens at the two extreme choices of λ in this case. For $\lambda = 1$, this becomes standard policy iteration, as the fixed point satisfies $\boldsymbol{v}^* = \mathscr{L}_{\pi_{n+1}} \boldsymbol{v}^*$ so that $\boldsymbol{v}^*$ must be the value of policy π_{n+1}. For $\lambda = 0$, one obtains standard value iteration, as the fixed point is reached under one step and is $\boldsymbol{v}^* = \mathscr{L}_{\pi_{n+1}} \boldsymbol{v}_n$, i.e., the approximate value of the one-step greedy policy. In general, the new value vector is moved only partially towards the direction of the Bellman update, depending on how we choose λ.

6.5.4.6 Linear Programming

Perhaps surprisingly, we can also solve an MDP through linear programming, reformulating the maximization problem as a linear optimization problem with linear constraints. As a first step we recall that there is an easy way to determine whether a particular $\boldsymbol{v}$ is an upper bound on the optimal value function $\boldsymbol{v}^*$, since if $\boldsymbol{v} \geq \mathscr{L}\boldsymbol{v}$ then $\boldsymbol{v} \geq \boldsymbol{v}^*$ by Theorem 6.5.3. In order to transform this into a linear program, we must first define a scalar function to minimize. We can do this by selecting some arbitrary distribution $\boldsymbol{y} \in \triangle(\mathcal{S})$ on the state space. Then denoting the vector of next state probabilities $p(s' \mid s, a)$ by $\boldsymbol{p}_{s,a}$, we can write the following linear program.

Primal linear program

$$\min_{\boldsymbol{v}} \boldsymbol{y}^\top \boldsymbol{v},$$

such that for all $s \in \mathcal{S}, a \in \mathcal{A}$

$$\boldsymbol{v}(s) - \gamma \boldsymbol{p}_{s,a}^\top \boldsymbol{v} \geq r(s).$$

Note that the inequality condition is equivalent to $\boldsymbol{v} \geq \mathscr{L}\boldsymbol{v}$, so that the smallest $\boldsymbol{v}$ that satisfies the inequality will be the optimal value function that satsifies the Bellman equation.

It also pays to look at the dual linear program, that aims to find the maximal cumulative discounted state-action visits $x(s, a)$ that are consistent with the transition kernel of the process.

Dual linear program

$$\max_{\boldsymbol{x}} \sum_{s \in \mathcal{S}} \sum_{a \in \mathcal{A}} x(s, a) \, r(s, a),$$

such that $x(s, a) \geq 0$ and for all $s' \in \mathcal{S}$

$$\sum_{a \in \mathcal{A}} x(s', a) - \sum_{s \in \mathcal{S}} \sum_{a \in \mathcal{A}} \gamma p(s' \mid s, a) \, x(s, a) = y(s')$$

with $\boldsymbol{y} \in \triangle(\mathcal{S})$.

In this case, the respective vector $\boldsymbol{x} \in \mathbb{R}^{|\mathcal{S} \times \mathcal{A}|}$ can be interpreted as the discounted sum of state-action visits, as shown in the following theorem.

Theorem 6.5.13 *For any policy π,*

$$x_\pi(s, a) = \mathbb{E}^\pi \left\{ \sum_{t \geq 0} \gamma^t \mathbb{I} \, s_t = s, a_t = a \mid s_0 \sim \boldsymbol{y} \right\}$$

is a feasible solution to the dual problem. On the other hand, if $\boldsymbol{x}$ is a feasible solution to the dual problem, then $\sum_a x(s, a) > 0$. Defining a respective randomized policy

$$\pi_{\boldsymbol{x}}(a \mid s) = \frac{x(s,a)}{\sum_{a' \in \mathcal{A}} x(s,a')},$$

$\boldsymbol{x}_{\pi_x} = \boldsymbol{x}$ *is a feasible solution of the dual program.*

The equality condition of the dual program ensures that $\boldsymbol{x}$ is consistent with the transition kernel of the MDP. Consequently, the program can be seen as search among all possible cumulative state-action distributions to find the one giving the highest total reward.

6.6 Optimality Criteria

While we have concentrated on discounted rewards under infinite horizon, we conclude this chapter with an overview of different optimality criteria. As already indicated previously, the following two views of discounted reward processes are equivalent.

Infinite horizon, discounted

Discount rewards by discount factor γ with $0 < \gamma < 1$. Then

$$U_t = \sum_{k=0}^{\infty} \gamma^k r_{t+k} \quad \text{and} \quad \mathbb{E}\, U_t = \sum_{k=0}^{\infty} \gamma^k \, \mathbb{E}\, r_{t+k}.$$

Geometric horizon, undiscounted

At each step t, the process terminates with probability $1 - \gamma$. That is,

$$U_t^T = \sum_{k=0}^{T-t} r_{t+k} \text{ with } T \sim \mathcal{G}eom(1-\gamma), \text{ so that } \mathbb{E}\, U_t = \sum_{k=0}^{\infty} \gamma^k \, \mathbb{E}\, r_{t+k}.$$

In the following, let Π be the set of all history-dependent (possibly randomized) policies. Similar to before, we write $\boldsymbol{v}_\gamma^\pi(s)$ for the total discounted reward of policy $\pi \in \Pi$ when starting in state s, now emphasizing the dependence on the discount factor γ.

Definition 6.6.1 π^* is *discount optimal* for $\gamma \in [0, 1)$ if

$$\boldsymbol{v}_\gamma^{\pi^*}(s) \geq \boldsymbol{v}_\gamma^\pi(s) \qquad \forall s \in \mathcal{S}, \pi \in \Pi.$$

The average reward (gain) criterion. Beside the expected total reward defined as $V_t^{\pi,T} \triangleq \mathbb{E}_\pi U_t^T$, the expected average reward is a natural criterion.

Definition 6.6.2 The gain g of a policy π starting from state s is defined as

$$g^\pi(s) \triangleq \lim_{T\to\infty} \frac{1}{T} V^{\pi,T}(s),$$

that is, the expected average reward the policy obtains when starting in s.

If the limit in Definition 6.6.2 does not exist, one may consider the limits

$$g_+^\pi(s) \triangleq \limsup_{T\to\infty} \frac{1}{T} V^{\pi,T}(s), \qquad g_-^\pi(s) \triangleq \liminf_{T\to\infty} \frac{1}{T} V^{\pi,T}(s).$$

Definition 6.6.3 π^* is *gain optimal* if

$$g^{\pi^*}(s) \geq g^\pi(s) \qquad \forall s \in \mathcal{S}, \pi \in \Pi.$$

Definition 6.6.4 π^* is *overtaking optimal* if

$$\liminf_{T\to\infty} \left[V^{\pi^*,T}(s) - V^{\pi,T}(s)\right] \geq 0 \qquad \forall s \in \mathcal{S}, \pi \in \Pi.$$

The existence of overtaking optimal policies is not guaranteed. The following criterion is a weaker one.

Definition 6.6.5 π^* is *average-overtaking optimal* if

$$\liminf_{T\to\infty} \frac{1}{T} \left[V^{\pi^*,T}(s) - V_+^\pi(s)\right] \geq 0 \qquad \forall s \in \mathcal{S}, \pi \in \Pi,$$

where $V_+^\pi(s) \triangleq \mathbb{E}_\pi \left(\sum_{t=1}^\infty \max\{r_t, 0\} \mid s_t = s\right)$.

Definition 6.6.6 π^* is *n-discount optimal* for $n \in \{-1, 0, 1, \ldots\}$ if

$$\liminf_{\gamma\uparrow 1} (1-\gamma)^{-n} \left[\boldsymbol{v}_\gamma^{\pi^*}(s) - \boldsymbol{v}_\gamma^\pi(s)\right] \geq 0 \qquad \forall s \in \mathcal{S}, \pi \in \Pi.$$

Definition 6.6.7 A policy is *Blackwell optimal* if $\forall s, \exists \gamma^*(s)$ such that

$$\boldsymbol{v}_\gamma^{\pi^*}(s) - \boldsymbol{v}_\gamma^\pi(s) \geq 0, \qquad \forall \pi \in \Pi, \gamma^*(s) \leq \gamma < 1.$$

Lemma 6.6.1 *If a policy is m-discount optimal then it is n-discount optimal for all $n \leq m$.*

Lemma 6.6.2 *Gain optimality is equivalent to (-1)-discount optimality.*

The different optimality criteria summarized here are treated in detail in Chapter 5 of [3].

6.7 Summary

Markov decision processes can be used to represent shortest path problems, stopping problems, experiment design problems, multi-armed bandit and more general reinforcement learning problems.

Bandit problems are the simplest type of Markov decision process, since they have a single fixed, never-changing state. However, to solve them, one can construct a Markov decision processes in belief space within a Bayesian framework. It is then possible to apply backwards induction to find the optimal policy.

Backwards induction is applicable more generally to arbitrary Markov decision processes. For the case of infinite-horizon problems, it is referred to as value iteration, as it converges to a fixed point. It is tractable when either the state space $\mathcal{S}$ or the horizon T are small (finite).

When the horizon is infinite, policy iteration can also be used to find optimal policies. It is different from value iteration in that at every step, it fully evaluates a policy before the improvement step, while value iteration only performs a partial evaluation. In fact, at the nth iteration, value iteration has calculated the value of an n-step policy.

We can arbitrarily mix between the two extremes of policy iteration and value iteration in two ways. Firstly, we can perform a k-step partial evaluation, which is called modified policy iteration. When $k = 1$ this coincides with value iteration, while for $k \to \infty$, one obtains policy iteration. Secondly, we can adjust our value function by using a temporal difference error of values in future time steps. Again, we can mix liberally between policy iteration and value iteration by focusing on errors far in the future (policy iteration) or on short-term errors (value iteration).

Finally, it is possible to solve MDPs through linear programming, reformulating the MDP as a linear optimization problem with constraints. In the primal formulation, we attempt to find a minimal upper bound on the optimal value function. In the dual formulation, our goal is to find a distribution on state-action visits that maximizes expected utility and is consistent with the MDP model.

6.8 Further Reading

For further information on the MDP formulation of bandit problems in the decision theoretic setting see the last chapter of [6], which was explored in more detail in Duff's PhD thesis [7]. When the number of (information) states in the bandit problem is finite, [8] has proven that it is possible to formulate simple so-called *index policies*. However, this is not generally applicable. Easily computable, near-optimal heuristic strategies for bandit problems will be presented in Chap. 10. The decision-theoretic solution to the unknown MDP problem is given in Chap. 9.

Further theoretical background on MDPs, including many of the theorems in Sect. 6.5, can be found in [3]. Chap. 2 of [9] gives a quick overview of MDP theory from the operator perspective. The introductory reinforcement learning book of [10] also explains the basic Markov decision process framework.

6.9 Exercises

6.9.1 MDP Theory

Exercise 6.1 Show that the expected discounted total reward of any given policy is equal to the expected undiscounted total reward with a finite, but random horizon T. In particular, let T be distributed according to a geometric distribution on $\{1, 2, \ldots\}$ with parameter $1 - \gamma$. Then show that

$$\mathbb{E} \lim_{T \to \infty} \sum_{k=0}^{T} \gamma^k r_k = \mathbb{E} \left(\sum_{k=0}^{T} r_k \,\middle|\, T \sim \mathcal{G}eom(1 - \gamma) \right).$$

6.9.2 Automatic Algorithm Selection

Consider the problem of selecting algorithms for finding solutions to a sequence of problems. Assume you have n algorithms to choose from. At time t, you get a task and choose one of the algorithms. Assume that the algorithms are randomized, so that each algorithm will find a solution with some unknown probability. Our aim is to maximize the expected total number of solutions found. Consider the following specific cases of this problem.

Exercise 6.2 Assume that the probability that the ith algorithm successfully solves the tth task is always p_i. Furthermore, tasks are in no way distinguishable from each other. In each case, let $p_i \in \{0.1, \ldots, 0.9\}$ and assume a prior distribution $\xi_i(p_i) = 1/9$ for all i with a complete belief $\xi(\boldsymbol{p}) = \prod_i \xi_i(p_i)$. Then formulate the problem as a decision-theoretic n-armed bandit problem with reward at time t being $r_t = 1$ if the task is solved and $r_t = 0$ if the problem is not solved. Independent of whether or not the task at time t is solved or not, at the next time-step the next problem is to be solved. The aim is to find a policy π mapping from the history of observations to the set of algorithms that maximizes the total reward to time T in expectation, i.e.,

$$\mathbb{E}_{\xi,\pi} U_0^T = E_{\xi,\pi} \sum_{t=1}^{T} r_t.$$

1. Characterize the essential difference between maximizing U_0^0, U_0^1, U_0^2.
2. For $n = 3$ algorithms, calculate the maximum expected utility

$$\max_{\pi} \mathbb{E}_{\xi,\pi} U_0^T$$

using backwards induction for $T \in \{0, 1, 2, 3, 4\}$ and report the expected utility in each case. *Hint: Use the decision-theoretic bandit formulation to dynamically construct a Markov decision process which you can solve with backwards induction. See also the extensive form of the utility from* (3.5.2).
3. Now utilize the backwards induction algorithm developed in the previous step in a problem where we receive a sequence of N tasks to solve and our utility is

$$U_0^N = \sum_{t=1}^{N} r_t.$$

At each step $t \leq N$, find the optimal action by calculating $\mathbb{E}_{\xi,\pi} U_t^{t+T}$ for $T \in \{0, 1, 2, 3, 4\}$. *Hint: At each step you can update your prior distribution using the same routine you use to update your prior distribution. You only need consider* $T < N - t$.
4. Develop a simple heuristic algorithm for your choice and compare its utility with the utility of backwards induction. Perform 10^3 simulations, each experiment running for $N = 10^3$ time-steps and average the results. How does the performance improve? *Hint: If the program runs too slowly go only up to* $T = 3$.

6.9.3 Scheduling

You are controlling a set of n processing nodes of a processing network that is part of a big CPU farm. At time t, you may be given a job of class $x_t \in X$ to execute. Assume these are identically and independently drawn such that $\mathbb{P}(x_t = k) = p_k$ for all t, k. With some probability p_0, you are not given a job to execute at the next step. If you do have a new job, then you can either:

(a) Ignore the job.
(b) Send the job to some node i. If the node is already active, then the previous job is lost.

Not all the nodes and jobs are equal. Some nodes are better at processing certain types of jobs. If the ith node is running a job of type $k \in X$, then it has a probability

$\phi_{i,k} \in [0, 1]$ of finishing it within that time step. Then the node becomes free and can accept a new job.

For this problem, assume that there are $n = 3$ nodes and $k = 2$ types of jobs and that the completion probabilities are given by the matrix

$$\boldsymbol{\Phi} = \begin{bmatrix} 0.3 & 0.1 \\ 0.2 & 0.2 \\ 0.1 & 0.3 \end{bmatrix}.$$

Further, we assume $p_0 = 0.1$, $p_1 = 0.4$, $p_2 = 0.5$ to be the probabilities of not getting any job, and the probabilities of the two job types, respectively. We wish to find the policy π maximizing the expected total reward given the MDP model μ, that is,

$$\mathbb{E}_{\mu,\pi} \sum_{t=0}^{\infty} \gamma^t r_t, \tag{6.9.1}$$

where $\gamma = 0.9$ and we get a reward of 1 every time a job is completed.

More precisely, at each time step t, the following events happen:

1. A new job x_t appears.
2. Each node either continues processing, or completes its current job and becomes free. You get a reward r_t equal to the number of nodes that complete their jobs within this step.
3. You decide whether to ignore the new job or add it to one of the nodes. If you add a job, then it immediately starts running for the duration of the time step. (If the job queue is empty then obviously you cannot add a job to a node.)

Exercise 6.3 Solve the following problems:

1. Identify the state and action space of this problem and formulate it as a Markov decision process. *Hint: Use independence of the nodes to construct the MDP parameters.*
2. Solve the problem using value iteration, using the stopping criterion indicated Theorem 6.5.8 (2c), where $\epsilon = 0.1$. Indicate the number of iterations needed to stop.
3. Solve the problem using policy iteration. Indicate the number of iterations needed to stop. *Hint: You can either modify the value iteration algorithm to perform policy evaluation, using the same ϵ, or you can use the linear formulation. If you use the latter, take care with the inverse!*
4. Now consider an alternative version of the problem, where we suffer a penalty of 0.1 (i.e., we get a negative reward) for each time-step that each node is busy. Are the solutions different?
5. Finally consider a version of the problem, where we suffer a penalty of 10 each time we cancel an executing job. Are the solutions different?
6. Plot the value function for the optimal policy in each setting.

Hint: To verify that your algorithms work, test them first on a smaller MDP with known solutions. For example, check out http://webdocs.cs.ualberta.ca/~sutton/book/ebook/node35.html.

6.9.4 General Questions

Exercise 6.4

1. What in your view are the fundamental advantages and disadvantages of modelling problems as Markov decision processes?
2. Is the algorithm selection problem of Exercise 6.2 solvable with policy iteration? If so, how?
3. What are the fundamental similarities and differences between the decision-theoretic finite-horizon bandit setting and the infinite-horizon MDP setting?

References

1. Chernoff, H.: Sequential design of experiments. Ann. Math. Stat. **30**(3), 755–770 (1959)
2. Chernoff, H.: Sequential models for clinical trials. In: Proceedings of the Fifth Berkeley Symposium on Mathematical Statistics and Probability, vol. 4, pp. 805–812. University of California Press (1966)
3. Puterman, M.L.: Markov Decision Processes: Discrete Stochastic Dynamic Programming. Wiley, New Jersey, US (1994)
4. Tseng, P.: Solving H-horizon, stationary Markov decision problems in time proportional to log(H). Oper. Res. Lett. **9**(5), 287–297 (1990)
5. Ye, Y.: The simplex and policy-iteration methods are strongly polynomial for the Markov decision problem with a fixed discount rate. Math. Oper. Res. **36**(4), 593–603 (2011)
6. DeGroot, M.H.: Optimal Statistical Decisions. Wiley (1970)
7. O'Gordon Duff, M.: Optimal learning computational procedures for bayes-adaptive markov decision processes. Ph.D. thesis, University of Massachusetts at Amherst (2002)
8. Gittins, J.C.: Multi-armed Bandit Allocation Indices. Wiley, New Jersey, US (1989)
9. Bertsekas, D.P., Tsitsiklis, J.N.: Neuro-Dynamic Programming. Athena Scientific (1996)
10. Sutton, R.S., Barto, A.G.: Reinforcement Learning: An Introduction. MIT Press (1998)

Chapter 7
Simulation-Based Algorithms

7.1 Introduction

In this chapter, we consider the general problem of reinforcement learning in dynamic environments. Up to now, we have only examined a solution method for bandit problems, which are only a special case. The Bayesian decision-theoretic solution as presented in Chap. 6 is to *reduce* the bandit problem to a *Markov decision process* which can then be solved with backwards induction.

On the other hand, we also have seen that MDPs can be used to *describe environments* in more general reinforcement learning problems. If the underlying MDP is fully known, then a number of standard algorithms—some of which we have introduced in Chap. 6—can be employed to find the optimal policy.

However, in the actual reinforcement learning problem, the model of the environment is *unknown*. Thus, the main focus of this chapter will be how to simultaneously learn about the underlying process and act to maximize utility in an *approximate* way. This can be done through approximate dynamic programming, where we replace the actual unknown parameters of the underlying process with estimates. The estimates can be improved by drawing samples from the environment, either by acting within the real environment or using a simulator. Although the algorithms following either approach may not be performing as well as the Bayes-optimal solution, they are computationally cheap so that they are worth investigating in practice.

It is important to note that while the algorithms presented in this chapter may converge eventually to an optimal policy, their *online* performance can be quite bad. That is, they may not accumulate a lot of reward while still learning. In that sense, they are not offering optimal online solutions to the reinforcement learning problem. For online algorithms with respective guarantees we refer to Chap. 10.

C. Dimitrakakis and R. Ortner, *Decision Making Under Uncertainty and Reinforcement Learning*, Intelligent Systems Reference Library 223,
https://doi.org/10.1007/978-3-031-07614-5_7

7.1.1 The Robbins-Monro Approximation

As already outlined, we are interested in reinforcement learning algorithms that replace the unknown actual process in which the learner is acting with an estimate that will eventually converge to the true process. Accordingly, the actions taken by the learner shall be nearly-optimal with respect to the estimate.

To approximate the process, one can use the general idea of Robbins-Monro stochastic approximation [1]. This entails maintaining a *point estimate* of the parameter we want to approximate and perform *random* steps that on average move towards the solution, in a way to be made more precise later. The stochastic approximation actually defines a large class of procedures, containing stochastic gradient descent as a special case.

Algorithm 7.1 shows an algorithm for the multi-armed bandit problem that uses a Robbins-Monro approximation. The parameters include a particular policy π which defines a probability distribution over the arms given the observed history, a set of initial estimates $\hat{r}_{i,0}$ for the mean of each arm i, and a sequence of step sizes $(\alpha_t)_t$.

The algorithm chooses an arm according to the policy π (line 3) after which the estimate $\hat{r}_{t,i}$ of the chosen arm i is updated (lines 4–6). As we shall see later, this particular update rule can be seen as trying to minimize the expected squared error between the estimated reward and the random reward obtained by each arm. Consequently, the variance of the reward of each arm plays an important role.

The step-sizes α_t are usually chosen to decrease with time t to guarantee convergence. Obviously, α_t can be chosen so that the $\hat{r}_{i,t}$ are simply the mean estimates of the expected value of the reward for each arm i, which is a natural choice.

Concerning an appriopriate choice of the policy used by the Robbins-Monro algorithm, note that all arms should be chosen often enough, so that one has good estimates for the expected reward of every arm. One simple way to achieve that is to play the apparently best arm most of the time, but to sometimes select an arm at random. This is called ϵ-greedy action selection.

Definition 7.1.1 (*ϵ-greedy policy*) Let $\hat{\mathcal{A}}_t^* = \arg\max_{i \in \mathcal{A}} \hat{r}_{t,i}$ denote the set of empirically best arms at step t. Then the ϵ-greedy policy $\hat{\pi}_\epsilon^*$ with parameters $(\epsilon_t)_t$ is given by

$$\hat{\pi}_\epsilon^* \triangleq (1 - \epsilon_t)\, \mathcal{Unif}(\hat{\mathcal{A}}_t^*) + \epsilon_t\, \mathcal{Unif}\,(\mathcal{A}).$$

In general it makes sense to let ϵ_t depend on t, so that the randomness can be chosen to decrease over time when estimates converge to their true values.

When using ϵ-greedy in our Robbins-Monro bandit algorithm, the two main parameters to choose are the amount of randomness ϵ_t and the step-size α_t in the estimation. Both of them have a significant effect on the performance of the algorithm. Although as indicated above, in general it makes sense to vary these parameters with time, it is perhaps instructive to look at what happens for fixed values of ϵ and α. Figure 7.1 shows the average reward obtained, if we keep the step size α or the randomness ϵ fixed, respectively, with initial estimates $\hat{r}_{0,i} = 0$.

Algorithm 7.1 Robbins-Monro bandit algorithm

input Set of arms $\mathcal{A} = \{1, \ldots, K\}$, step-sizes $(\alpha_t)_t$, initial estimates $(\hat{r}_{i,0})_i$, policy π.
for $t = 1, \ldots, T$ **do**
  Choose arm $a_t = i$ with probability $\pi(i \mid a_1, \ldots, a_{t-1}, r_1, \ldots, r_{t-1})$.
  Observe reward r_t.
  $\hat{r}_{t,i} = \alpha_t r_t + (1 - \alpha_t)\hat{r}_{i,t-1}$ `// estimation step`
  $\hat{r}_{t,i} = \hat{r}_{j,t-1}$ for $j \neq i$
end for

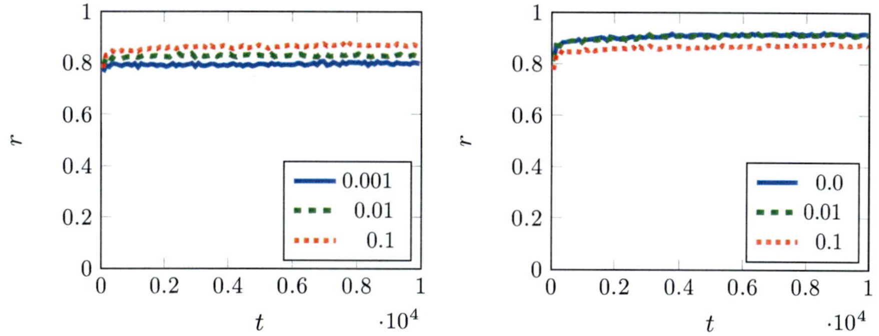

(a) fixed ϵ (b) fixed α

Fig. 7.1 The plots above show average reward over time. For fixed $\epsilon_t = 0.1$, the step size is $\alpha \in \{0.01, 0.1, 0.5\}$. For fixed step size $\alpha = 0.1$ we vary the exploration rate in $\epsilon \in \{0.0, 0.01, 0.1\}$

For fixed ϵ, we find that larger values of α tend to give better results eventually, while smaller values have a better initial performance. This is a natural trade-off, since large α appears to "learn" fast, but it also "forgets" quickly. That is, for a large α the estimates mostly depend upon the last few rewards that have been observed.

Things are not so clear-cut for the choice of ϵ. We see that the choice of $\epsilon = 0$ is significantly worse than $\epsilon = 0.1$. That appears to suggest that there is an optimal level of exploration. Ideally, we should use the decision-theoretic solution seen in Chap. 6, but usually a good heuristic method for choosing ϵ will be good enough.

7.1.2 *The Theory of the Approximation*

An important question when working with approximation algorithms as the Robbins-Monro bandit algorithm of the previous section is whether the used estimates converge to the correct values, so that the algorithm itself eventually converges to an optimal policy. Note that as already mentioned we want to focus on asymptotic convergence and are not interested in how much reward we obtain in the learning phase.

This section briefly reviews some basic results of stochastic approximation theory. Complete proofs can be found in [2]. We first consider the core problem of stochastic approximation itself. In particular, we shall cast the approximation problem as a minimization problem. That is, given some unknown parameter of the environment $\nu \in \mathbb{R}^n$ and an estimate $\nu_t \in \mathbb{R}^n$ of ν at time t, the estimate is assumed to be chosen such that for a certain function $f : \mathbb{R}^n \to \mathbb{R}$ the estimate ν_t minimizes f.

Then with respect to the learning algorithm the goal is that the sequence of values $(\nu_t)_t$ generated by the algorithm converges to some ν^* that is a local minimum, or a stationary point for f. For strictly convex f, this would also be a global minimum.

In the following, we will examine algorithms that compute estimates ν_t according to the update

$$\nu_{t+1} = \nu_t + \alpha_t z_{t+1}, \tag{7.1.1}$$

where $\alpha_t \in \mathbb{R}$ is a step-size and $z_t \in \mathbb{R}^n$ a random direction. This process can be shown to converge to a stationary point of f under certain assumptions. Sufficient conditions include continuity and smoothness properties of f and the update directions z_t. In particular, we shall assume the following about the function f that we wish to minimize.

Assumption 7.1.1 Let h_t denote that history $(\mu_t, z_t, \alpha_t, \ldots, \mu_0, z_0, \alpha_0)$ up to step t. Then function $f : \mathbb{R}^n \to \mathbb{R}$ satisfies:

(i) $f(x) \geq 0$ for all $x \in \mathbb{R}^n$.
(ii) (*Lipschitz derivative*) f is continuously differentiable (i.e., the derivative ∇f exists and is continuous) and $\exists L > 0$ such that

$$\|\nabla f(x) - \nabla f(y)\| \leq L \|x - y\|, \qquad \forall\, x, y \in \mathbb{R}^n.$$

(iii) (*Pseudo-gradient*) $\exists c > 0$ such that

$$c \|\nabla f(\nu_t)\|^2 \leq -\nabla f(\nu_t)^\top \mathbb{E}(z_{t+1} \mid h_t), \qquad \forall\, t.$$

(iv) $\exists K_1, K_2 > 0$ such that

$$\mathbb{E}(\|z_{t+1}\|^2 \mid h_t) \leq K_1 + K_2 \|\nabla f(\mu_t)\|^2, \qquad \forall\, t.$$

The basic condition (ii) ensures that the function is well-behaved, so that gradient-following methods will easily find the minimum. Condition (iii) combines two assumptions in one. Firstly, the expected direction of the update always decreases

f, and secondly that the squared norm of the gradient is not too large relative to the size of the update. Finally, condition (iv) ensures that the update is bounded in expectation relative to the gradient. One can see how putting together the last two conditions ensures that the expected direction of the update is correct and its norm is bounded.

Theorem 7.1.1 *Consider an algorithm with update* (7.1.1) *such that the step sizes* $\alpha_t \geq 0$ *satisfy*

$$\sum_{t=0}^{\infty} \alpha_t = \infty, \qquad \sum_{t=0}^{\infty} \alpha_t^2 < \infty.$$

Then under Assumption 7.1.1 the following holds with probability 1:

1. *The sequence* $\left(f(\nu_t)\right)_t$ *converges.*
2. $\lim_{t\to\infty} \nabla f(\nu_t) = 0$.
3. *Every limit point* ν^* *of* ν_t *satisfies* $\nabla f(\nu^*) = 0$.

The conditions of Theorem 7.1.1 are not necessary. Alternative sufficient conditions relying on contraction properties are for example discussed in detail in [2]. The following example illustrates the impact of the choice of step size schedule $(\alpha_t)_t$ on convergence.

Example 7.1 (Estimating the mean of a Gaussian distribution) Consider a sequence of observations x_t sampled from a Gaussian distribution with mean $\frac{1}{2}$ and variance 1, that is, $x_t \sim \mathcal{N}(0.5, 1)$. We try to estimate the mean using updates according to (7.1.1) with the update direction given by

$$z_{t+1} = x_{t+1} - \nu_t.$$

Now consider three different step-size schedules: The first one, $\alpha_t = 1/t$, satisfies both assumptions of Theorem 7.1.1 on the step-size. The second one, $\alpha_t = 1/\sqrt{t}$, decreases too slowly, while the third one, $\alpha_t = t^{-3/2}$, approaches zero too fast.

Figure 7.2 demonstrates the convergence, or lack thereof, of the estimates ν_t to the expected value. In fact, the schedule $t^{-3/2}$ converges to a value quite far away from the expected value, while the slow schedule $1/\sqrt{t}$ oscillates.

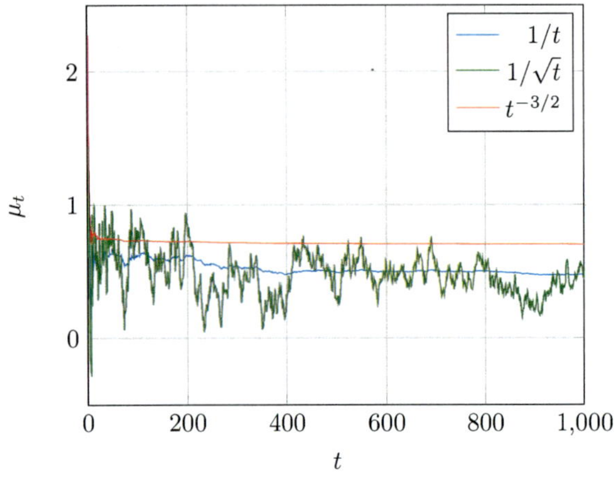

Fig. 7.2 Estimation of the expectation of $x_t \sim \mathcal{N}(0.5, 1)$ using use three step-size schedules

7.2 Dynamic Problems

It is possible to extend the ideas outlined in the previous section to the dynamic Markov decision process setting. We simply need to have a policy that is greedy with respect to our estimates as well as a way to update our estimates so that they converge to the actual MDP we are acting in. The additional challenge of the dynamic setting is that our policy now affects which sequences of states we observe. That is, while in the bandit problem we could freely select an arm to pull, sampling an arbitrary state is not possible as easily (unless some simulation of the MDP is available). Otherwise, the algorithmic structure remains basically the same and is described in Algorithm 7.2.

Algorithm 7.2 Generic reinforcement learning algorithm

1: **input** state space $\mathcal{S}$, action space $\mathcal{A}$, reward set $\mathcal{R} \subseteq \mathbb{R}$, parameter set Θ, initial parameters $\theta_0 \in \Theta$, update-rule $f : \Theta \times \mathcal{S}^2 \times \mathcal{A} \times \mathcal{R} \to \Theta$, policy $\pi : \mathcal{S} \times \Theta \to \triangle(\mathcal{A})$
2: **for** $t = 1, \ldots, T$ **do**
3: Take action $a_t \sim \pi(\cdot \mid \theta_t, s_t)$.
4: Observe reward r_{t+1}, state s_{t+1}.
5: Update estimate $\theta_{t+1} = f(\theta_t, s_t, a_t, r_{t+1}, s_{t+1})$.
6: **end for**

In general, the policy chooses an action a_t according to a distribution that depends on the current state s_t and a parameter θ_t that e.g. could describe a posterior distribution over MDPs, or a distribution over other parameters. Concerning the choice of policy one can e.g. use the Bayes-optimal policy with respect to the parameter θ or some heuristic policy.

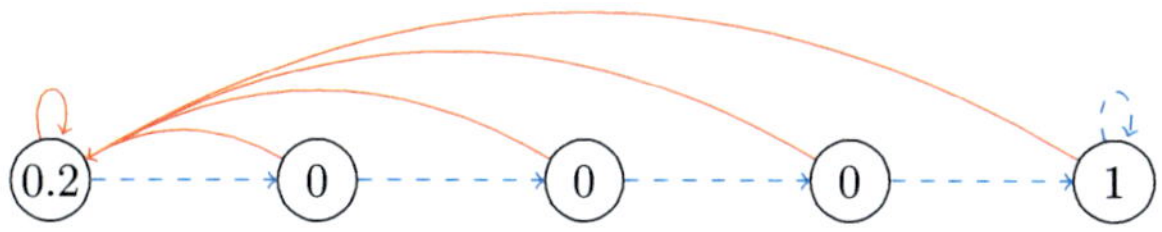

Fig. 7.3 The chain task

Table 7.1 The chain task's value function for $\gamma = 0.95$

s	s_1	s_2	s_3	s_4	s_5
$V^*(s)$	7.6324	7.8714	8.4490	9.2090	10.209
$Q^*(s, 1)$	7.4962	7.4060	7.5504	7.7404	8.7404
$Q^*(s, 2)$	7.6324	7.8714	8.4490	9.2090	10.2090

Example 7.2 (The chain task) The chain task has two actions and five states, as shown in Fig. 7.3. The starting state is the leftmost state, where the reward is 0.2. The reward is 1.0 in the rightmost state, and zero otherwise. The first action (dashed, blue) takes you to the right, while the second action (solid, red) takes you to the leftmost state, both with probability 0.8. With a probability of 0.2 both actions have the opposite effect. The value function of the chain task for a discount factor $\gamma = 0.95$ starting in the leftmost state is shown in Table 7.1.

The chain task is a very simple but well-known task used to test the efficacy of reinforcement learning algorithms. In particular, it is useful for analysing how algorithms solve the exploration-exploitation trade-off, since in the short run (i.e., small discount factor γ) simply staying in the leftmost state is advantageous. However if the discount factor is sufficiently large, algorithms should be incentivized to more fully explore the state space. A variant of this task with action-dependent rewards can be found in [3].

7.2.1 Monte Carlo Policy Evaluation and Iteration

To make things as easy as possible, let us assume that we have a way to start the environment from any arbitrary state. That would be the case if the environment had a *reset action*, or if we were simply running an accurate simulation.

We shall begin with simplest possible problem, that of estimating the expected utility of each state for a specific policy. This can be performed with Monte Carlo policy evaluation as shown in Algorithm 7.3. The idea is to estimate the value function for every state by approximating the expectation with the sum of rewards obtained over multiple trajectories starting from each state. Although this is very simple, Monte Carlo policy evaluation is a very useful method that is employed by several more complex algorithms as a subroutine. It can be also used non-Markovian environments such as partially observable or multi-agent settings.

Algorithm 7.3 Monte Carlo policy evaluation

input policy π, number of iterations K, horizon T
for $s \in \mathcal{S}$ **do**
  for $k = 0, \ldots, K$ **do**
    Initialize $U^{(k)} := 0$.
    Choose initial state $s_1 = s$.
    for $t = 0, \ldots, T$ **do**
      Choose action $a_t \sim \pi(\cdot \mid h_t)$ for history h_t.
      Observe reward r_t and next state s_{t+1}.
      Set $U^{(k)} := U^{(k)} + r_t$.
    end for
  end for
  Calculate estimate $\boldsymbol{v}_K(s) = \frac{1}{K}\sum_{k=1}^{K} U^{(k)}$.
end for
return $\boldsymbol{v}_K$

Remark 7.2.1 Using Hoeffding's inequality (4.5.5), one can show that for each state s that the error of the estimate $\boldsymbol{v}_K(s)$ is bounded by $\sqrt{\frac{\ln(2|\mathcal{S}|/\delta)}{2K}}$ with probability at least $1 - \frac{\delta}{|\mathcal{S}|}$. Hence, a union bound of the form $P(A_1 \cup A_2 \cup \ldots \cup A_n) \leq \sum_i P(A_i)$ shows that with probability $1 - \delta$ this error bound holds for all states.

Stochastic policy evaluation as shown in Algorithm 7.4 is a generalization of Monte Carlo policy evaluation. Choosing the parameter $\alpha_k = \frac{1}{k}$ and initial values $\boldsymbol{v}(s) = 0$ the two algorithms are the same.

Algorithm 7.4 Stochastic policy evaluation

input policy π, initial values $\boldsymbol{v}_0$, parameters α_k, number of iterations K, horizon T
for $s \in \mathcal{S}$ **do**
  for $k = 1, \ldots, K$ **do**
    Choose initial state $s_1 = s$.
    for $t = 0, \ldots, T$ **do**
      Choose action $a_t \sim \pi(\cdot \mid h_t)$ for history h_t.
      Observe reward r_t and next state s_{t+1}.
      Set $U^{(k)} := U^{(k)} + r_t$.
    end for
    Update estimate $\boldsymbol{v}_k(s) = \boldsymbol{v}_{k-1}(s) + \alpha_k(U^{(k)} - \boldsymbol{v}_{k-1}(s))$.
  end for
end for
return $\boldsymbol{v}_K$

7.2.2 *Monte Carlo Updates*

It is also possible to update the value of all encountered states after each visit. This algorithm is called *every-visit Monte Carlo*, whose evaluation of a given state trajectory $s_1, \ldots, s_T$ and thereby earned rewards $r_1, \ldots, r_T$ is shown in Algorithm 7.5. In general, an estimate of the value function will be computed over several iterations as in the algorithms of the previous section. The parameters α_k we had before are here replaced by the values $\frac{1}{n_t(s)}$ that depend on the state and the respective visits in the latter. However, the type of estimate computed by every-visit Monte Carlo can be

Algorithm 7.5 Every-visit Monte Carlo update

input initial values $\boldsymbol{v}$, state visit counts $n(s)$, trajectory $s_1, \ldots, s_T$, rewards $r_1, \ldots, r_T$
Initialize $U := 0$.
for $t = 1, \ldots, T$ **do**
    $U = U + r_t$.
    $n(s_t) = n(s_t) + 1$
    $\boldsymbol{v}(s_t) = \boldsymbol{v}(s_t) + \frac{1}{n_t(s_t)}(U_t - \boldsymbol{v}(s_t))$
end for
return $\boldsymbol{v}$

biased, as the update is going to be proportional to the number of steps spent in the respective state. In order to avoid the bias, we can instead only perform the update on each first visit to every state. This eliminates the dependence between states and is called the *first-visit Monte Carlo* algorithm.

Algorithm 7.6 First-visit Monte Carlo update

input Initial values $\boldsymbol{v}$, state visit counts $n(s)$, trajectory $s_1, \ldots, s_T$, rewards $r_1, \ldots, r_T$
Initialize $U := 0$.
for $t = 1, \ldots, T$ **do**
    $U = U + r_t$
    if $s_t \notin \{s_1, \ldots, s_{t-1}\}$ **then**
        $n(s_t) = n(s_t) + 1$
        $\boldsymbol{v}(s_t) = \boldsymbol{v}(s_t) + \frac{1}{n_t(s_t)}(U_t - \boldsymbol{v}(s_t))$
    end if
end for
return $\boldsymbol{v}$

Figure 7.4 shows the difference between the two Monte Carlo evaluation methods on the chain task for the optimal policy. In particualr, it shows the L1 error between the value function of the optimal policy and the respective Monte Carlo estimates.

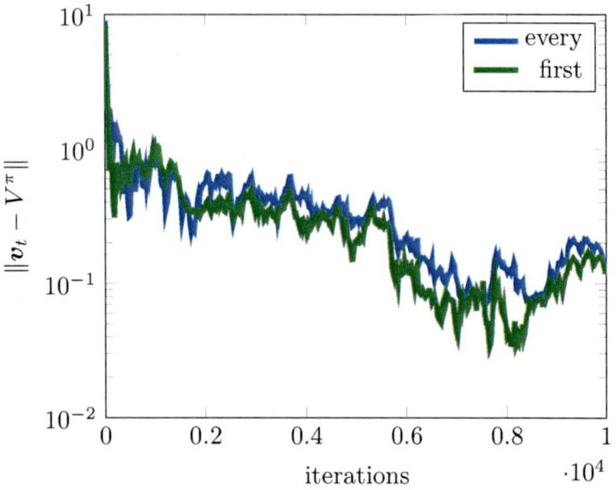

Fig. 7.4 Error as the number of iterations n increases, for first and every visit Monte Carlo estimation

7.2.3 *Temporal Difference Methods*

The kind of update we have now seen in several algorithms is of the form

$$\boldsymbol{v}(s_t) = \boldsymbol{v}(s_t) + \alpha(U_t - \boldsymbol{v}(s_t)), \tag{7.2.1}$$

where U_t is the utility sampled along some trajectory that was generated when following policy π. The difference between the observed utility and the assumed value so far can be seen as an error term, which is used for correction towards the true value.

The main idea of temporal difference learning methods is to replace the utility term by the immediate reward(s) in the next step(s) and estimate the utility for the subsequent steps by the current value of the next state.

That is, in the simplest case (now for technical reasons considering discounted rewards) the error considered is the so-called *temporal difference! error*

temporal difference! error

$$d_t = r_t + \gamma \boldsymbol{v}(s_{t+1}) - \boldsymbol{v}(s_t), \tag{7.2.2}$$

so that the update is

$$\boldsymbol{v}(s_t) = \boldsymbol{v}(s_t) + \alpha d_t.$$

It can be shown that the error term used in (7.2.1) can be written as a sum over the temporal difference errors

$$U_t - \boldsymbol{v}(s_t) = \sum_{\ell \geq t} \gamma^{\ell - t} d_\ell,$$

provided that $\boldsymbol{v}$ is assumed to be fixed. Otherwise the sum at least provides an approximation.

7.2.3.1 Temporal Difference Algorithm with Eligibility Traces

Instead of the estimate $r_t + \gamma \boldsymbol{v}(s_{t+1})$ we used for the utility before we can more generally consider the return over the next m steps starting in step t, that is, $U_{t,m} := r_t + \gamma r_{t+1} + \ldots + \gamma^{m-1} r_{t+m-1} + \gamma^m \boldsymbol{v}(s_{t+m})$. The TD($\lambda$) algorithm now uses a mixture

$$U_t^\lambda := (1 - \lambda) \sum_{m \geq 1} \lambda^{m-1} U_{t,m}$$

of all these different estimates, where the parameter $\lambda \in [0, 1]$ determines the importance of estimates with higher m similar to a discount factor and the factor $(1 - \lambda)$ is used for normalization. Note that for $\lambda = 0$ one obtains the estimate corresponding to using the temporal difference error (7.2.2). If there is a terminal state then for $\lambda = 1$ an update with respect to U_t^λ can be shown to correspond to a Monte Carlo update. In general, doing the update according to (7.2.1) with respect to U_t^λ, one can write the update using the temporal difference error as follows.

TD(λ) update

$$\boldsymbol{v}(s_t) = \boldsymbol{v}(s_t) + \alpha \sum_{\ell=t}^{\infty} (\gamma\lambda)^{\ell-t} d_\ell .$$

Unfortunately, as this update uses all future state and reward observations, it is only possible to implement it offline. However, considering a so-called *backward view* obtaining an online version shown as Algorithm 7.7 is possible.

Algorithm 7.7 Online TD(λ)

---

input initial values, trajectory $s_1, \ldots, s_T$, rewards $r_1, \ldots, r_T$
$\boldsymbol{e} = \boldsymbol{0}$
for $t = 1, \ldots, T$ **do**
  $\boldsymbol{e}(s_t) = \boldsymbol{e}(s_t) + 1$ `// eligibility increase`
  for $s \in \mathcal{S}$ **do**
    $\boldsymbol{v}(s) = \boldsymbol{v}(s) + \alpha \boldsymbol{e}(s)\, d_t$ `// update eligible states`
  end for
  $\boldsymbol{e} = \lambda \boldsymbol{e}_t$ `// eligibility discount`
end for
return $\boldsymbol{v}$

The main idea is to backpropagate changes in future states to previously encountered states. Thereby we wish to modify older states less than more recent states. This is accomplished by using the *eligibility traces* $\boldsymbol{e}$. These are initialized to be 0 for all states. Each visit increases the eligibility of a state by 1, and there is a general discount of λ for all states in each step. Thus, the eligibility takes into account how often states

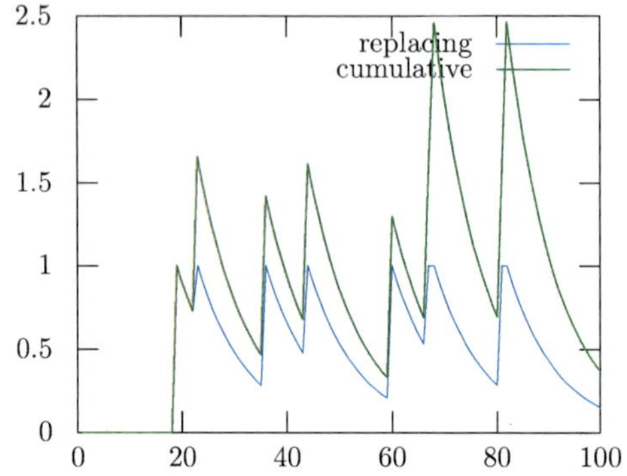

Fig. 7.5 Eligibility traces, replacing and cumulative

are visited and discards visits that are not so recent. Accordingly, the eligibility of each state is used as a factor for the temporal difference error in the state updates.

Figure 7.5 is an illustration of how eligibility traces operate in the two different formulations: replacing versus cumulative. While replacing traces start from 1 whenever a state is visited again, cumulative traces increase with every visit.

7.2.4 *Stochastic Value Iteration Methods*

In the previous sections we have seen the idea of updating value information not on all states but only on the currently observed one. This approach can not only be applied to the setting when one wants to estimate the value of a given policy, but also when we want to learn an optimal policy. For example, standard value iteration as introduced in Sect. 6.5.4.1 performs a sweep over the complete state space at each iteration. However, in principle one could perform the update over an arbitrary sequence of states. These could be generated when following a particular fixed policy or more generally a reinforcement learning algorithm. This lends to the idea of *simulation-based* value iteration.

In order to work in general the used state sequences usually must satisfy certain technical requirements. One sufficient condition is for example that the policies that generate the state sequences must be *proper* for episodic problems with a terminal state. That is, they all reach a terminating state with probability 1. For discounted non-episodic problems, this can be achieved by using a geometric distribution for a restart in a state chosen uniformly at random. This ensures that all policies will be proper. Alternatively, one could also select starting states with an arbitrary schedule, as long as it is guaranteed that all states are visited infinitely often in the limit.

7.2.4.1 Simulation-Based Value Iteration

Before considering the general case when the underlying MDP model μ is unknown, we start with the obviously simpler case where we can obtain data of the MDP from simulation. That is, for each state-action pair (s, a), we can request independent samples from the respective transition probability distribution $p_\mu(\cdot|s, a)$. Algorithm 7.8 shows a generic simulation-based value iteration algorithm. In this algorithm, at every step there is a probability ϵ the agent moves to a random state. This can be seen as restarting the MDP from a uniform starting state distribution $\mathcal{Unif}(\mathcal{S})$.

Algorithm 7.8 Simulation-based value iteration

```
1: input μ-simulator
2: Initialize s_1 ∈ S, v_0 ∈ V.
3: for t = 1, 2, ..., T do
4:   π_t(s_t) = r_μ(s) + argmax_a γ Σ_{s'∈S} p_μ(s'|s_t, a) v_{t-1}(s')
5:   v_t(s_t) = r_μ(s) + γ Σ_{s'∈S} p_μ(s'|s_t, π_t(s_t)) v_{t-1}(s')
6:   s_{t+1} ~ (1 − ε) · p_μ(s_{t+1} | s_t, π_t(s_t)) + ε · Unif(S)
7: end for
8: return π_T, V_T
```

Figure 7.6 shows the error in value function estimation in the chain task (Example 7.2) when using simulation-based value iteration. In general, it is advisable to use an initial value $\boldsymbol{v}_0$ that is an upper bound on the optimal value function, if such a value is known. This is due to the fact that in that case convergence of simulation-based value iteration is always guaranteed, as long as the policy that we are using is proper.[1] This is confirmed by the results. In general, the estimation error is highly dependent upon the initial value function estimate $\boldsymbol{v}_0$ and the exploration parameter ϵ. It is interesting to see uniform sweeps (i.e., $\epsilon = 1$) result in the lowest estimation error.

7.2.4.2 *Q*-Learning

Simulation-based value iteration can be suitably modified for the general reinforcement learning setting when neither model nor simulator for the underlying environment are available. Here instead of relying on a model of the environment, we replace arbitrary random sweeps of the state-space with the actual state sequence observed in the real environment. We also use this sequence as a simple way to estimate the transition probabilities. The replacement of the true MDP model by an estimate is a natural idea which leads to a whole family of stochastic value iteration algorithms. The most important of these is Q-learning, which uses a trivial empirical MDP model.

[1] As mentioned, in the case of discounted non-episodic problems, this amounts to a geometric stopping time distribution, after which the state is drawn from the initial state distribution.

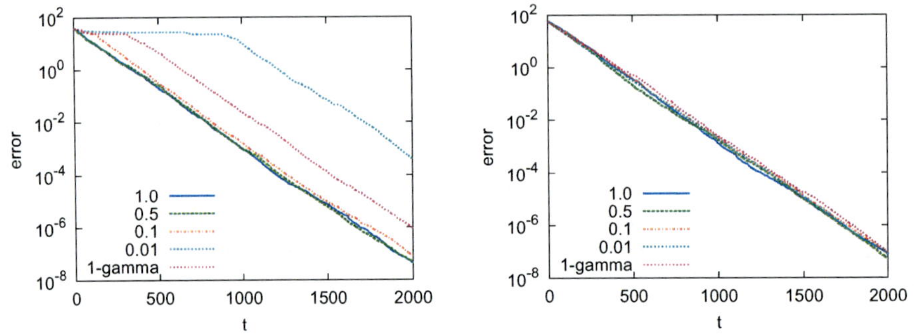

(a) Pessimistic initialization. (b) Optimistic initialization.

Fig. 7.6 Simulation-based value iteration with pessimistic initial estimates ($\boldsymbol{v}_0 = 0$) and optimistic initial estimates ($\boldsymbol{v}_0 = 20 = 1/(1-\gamma)$) for varying ϵ. Errors indicate $\|\boldsymbol{v}_n - V^*\|_1$

Q-learning is shown as Algorithm 7.9, not only one of the most well-known but also one of the simplest algorithms in reinforcement learning.

Algorithm 7.9 Q-learning

```
input exploration parameters ε_t, step sizes α_t
Initialize s_1 ∈ S, v_0 ∈ V.
for t = 1, 2, ..., T do
  v_t(s) = max_{a∈A} q_t(s, a)                       // update value function
  π̂*_t(s) = arg max_{a∈A} q_t(s, a)            // compute optimal policy wrt v
  Choose a_t ~ π̂*_{ε_t}(s_t).                       // play ε_t-greedy policy
  q_{t+1}(s_t, a_t) = (1 − α_t) q_t(s_t, a_t) + α_t[r_μ(s_t) + v_t(s_{t+1})]
  Observe s_{t+1} ~ p_μ(s_{t+1} | s_t, a_t).
end for
return π_T, V_T
```

It is straightforward to verify that Q-learning is just a version of simulation-based value iteration, where at each step t, given the partial observation (s_t, a_t, s_{t+1}) the transition model of the underlying MDP is approximated by

$$\hat{p}(s'|s_t, a_t) = \begin{cases} 1, & \text{if } s_{t+1} = s' \\ 0, & \text{if } s_{t+1} \neq s'. \end{cases}$$

In addition, since we cannot arbitrarily select states in the real environment, we replace the state-exploring parameter ϵ with a time-dependent exploration parameter ϵ_t for the policy we employ on the real environment. Even though this model is very simplistic, it still works relatively well in practice.

Figure 7.7 shows the performance of the basic Q-learning algorithm for the chain task (Example 7.2) in terms of value function error and regret. In this particular implementation, we used a polynomially decreasing exploration parameter ϵ_t and

Fig. 7.7 Q-learning with $v_0 = 1/(1-\gamma)$, $\epsilon_t = 1/n(s_t)$, $\alpha_t = \alpha n(s_t)^{-2/3}$ for state-visit counts $n(s)$

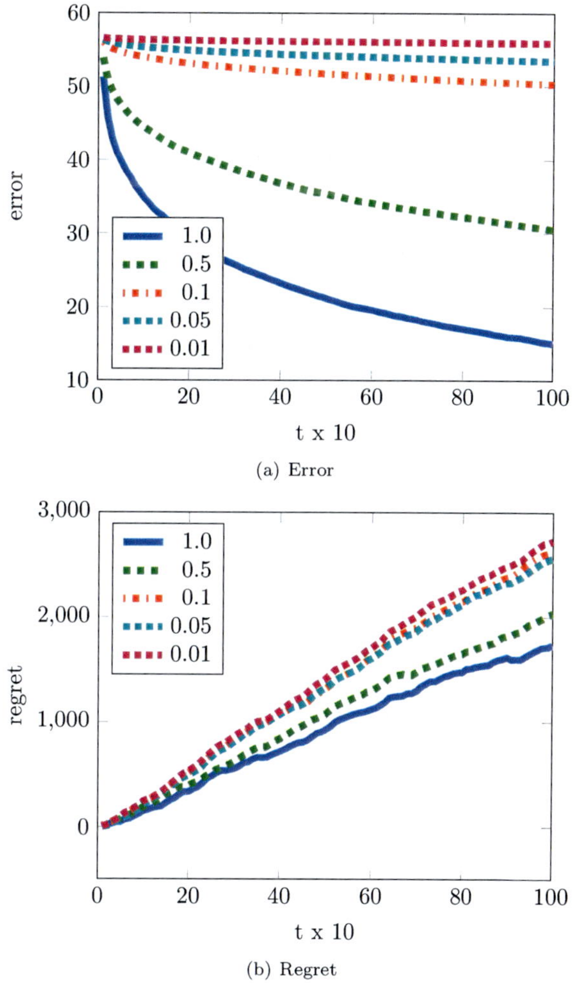

step size α_t. Both of these depend on the number of visits to a particular state, which leads to a more efficient performance.

Of course, one could get any algorithm in between pure Q-learning and pure stochastic value iteration. In fact, variants of the Q-learning algorithm using eligibility traces (see Sect. 7.2.3.1) can be formulated in the same way.

7.2.4.3 Generalized Stochastic Value Iteration

Finally, we present an algorithm, which can be considered as a generalization of the algorithms we have seen In particular, generalized stochastic value iteration as shown as Algorithm 7.10 is an online reinforcement learning algorithm, that includes simulation-based value iteration and Q-learning as special cases. As Q-learning and other algorithms we have seen before there are parameters ϵ_t for the amount of exploration as well as the step size parameters α_t. In addition the algorithm uses state-action distributions σ_t and an initial MDP estimator $\hat{\mu}_0$. The MDP estimators $\hat{\mu}$ used and updated by the algorithm are supposed to include both an estimate of the transition probabilities $p_{\hat{\mu}}(s' \mid s, a)$ and the expected reward[2] $r_{\hat{\mu}}(s)$.

Algorithm 7.10 Generalized stochastic value iteration

```
input exploration parameters ε_t, step sizes α_t, state-action distributions σ_t, initial MDP estimate μ̂_0
Initialize s_1 ∈ S, q_1 ∈ Q, v_0 ∈ V.
for t = 1, 2, ..., T do
    v_t(s) = max_{a∈A} q_t(s, a)
    π̂*_t(s) = arg max_{a∈A} q_t(s, a)
    Choose a_t ~ π̂*_{ε_t}(s_t).
    Observe s_{t+1}, r_{t+1}.
    Update MDP estimate μ̂_t = μ̂_{t−1} | s_t, a_t, s_{t+1}, r_{t+1}.
    for s ∈ S, a ∈ A do
        With probability σ_t(s, a) update
```

$$\boldsymbol{q}_{t+1}(s, a) = (1 - \alpha_t)\, \boldsymbol{q}_t(s, a) + \alpha_t \left[r_{\hat{\mu}_t}(s) + \gamma \sum_{s' \in \mathcal{S}} p_{\hat{\mu}_t}(s' \mid s, a)\, \boldsymbol{v}_t(s') \right],$$

```
        otherwise set q_{t+1}(s, a) = q_t(s, a).
    end for
end for
return π_T, V_T
```

It is instructive to examine special cases for these parameters. When $\sigma_t = 1$, $\alpha_t = 1$, and $\widehat{\mu_t} = \mu$, we obtain standard value iteration. Q-learning is obtained when setting $\sigma_t(s, a) = \mathbb{I}\{s_t = s, a_t = a\}$ and

$$p_{\hat{\mu}_t}(s_{t+1} = s' \mid s_t = s, a_t = a) = \mathbb{I}\left\{s_{t+1} = s' \mid s_t = s, a_t = a\right\}.$$

Finally, for $\sigma_t(s, a) = \boldsymbol{e}_t(s, a)$ we obtain a stochastic eligibility-trace Q-learning algorithm similar to TD(λ).

[2] More generally, rewards could of course depend not only on state, but the combination of state and action.

7.3 Discussion

Most of the algorithms we have seen in this chapter are quite simple and thus clearly demonstrate the principle of learning by reinforcement. However, they do not aim to solve the reinforcement learning problem in an optimal way. They have been mostly used for finding near-optimal policies given access to samples from a simulator, for example in the case of learning to play Atari games [4]. However, even in this case, a crucial issue is how much data is needed to be able to approach optimal behavior.

Convergence. Even though it is quite simple, Q-learning can be shown to converge to an optimal policy in various settings. Tsitsiklis [5] has provided an asymptotic proof based on stochastic approximation theory with less restrictive assumptions than the original paper of [6]. Later [7] proved finite sample convergence results under strong mixing assumptions on the MDP. Moreover, for modifications such as delayed Q-learning [8] stronger results on the sample complexity can be shown. That is, $\tilde{O}(|\mathcal{S}||\mathcal{A}|)$ samples are sufficient to determine an ϵ-optimal policy with high probability.

Exploration. Another extension of Q-learning is using a population of value function estimates using random initial values and weighted bootstrapping. This was first introduced by [9, 10] and evaluated on bandit tasks. Recently, this idea has also been exploited in the context of deep neural networks by [11] for representations of value functions in the setting of full reinforcement learning. We will examine this more closely in Chap. 8.

Bootstrapping and subsampling use a single set of empirical data to obtain an empirical measure of uncertainty about statistics of the data. We wish to do the same thing for value functions, based on data from one or more trajectories. Informally, this variant maintains a collection of Q-value estimates, each one of which is trained on different segments[3] of the data, with possible overlaps. In order to achieve efficient exploration, a random Q-estimate is selected at every episode or every few steps. This results in a bootstrap analogue of Thompson sampling. Figure 7.8 shows the use of weighted bootstrap estimates for the Double Chain problem introduced by [3].

7.4 Exercises

Exercise 7.1 According to Theorem 6.5.1 the value function of a policy π for an MDP μ with state reward function $\boldsymbol{r}$ can be written as the solution of the linear equation $\boldsymbol{v}_\mu^\pi = (I - \gamma P_\mu^\pi)^{-1}\boldsymbol{r}$, where the term $\Phi_\mu^\pi \triangleq (I - \gamma P_\mu^\pi)^{-1}$ can be seen as a feature matrix. However, simulation-based algorithms like Sarsa only approximate the value function directly rather than Φ_μ^π. This means that if the reward function changes, they have to be restarted from scratch. Is there a way to improve on this?[4]

[3] If not dealing with bandit problems, it is important to do this with trajectories.

[4] This exercise stems from a discussion with Peter Auer in 2012 about this problem.

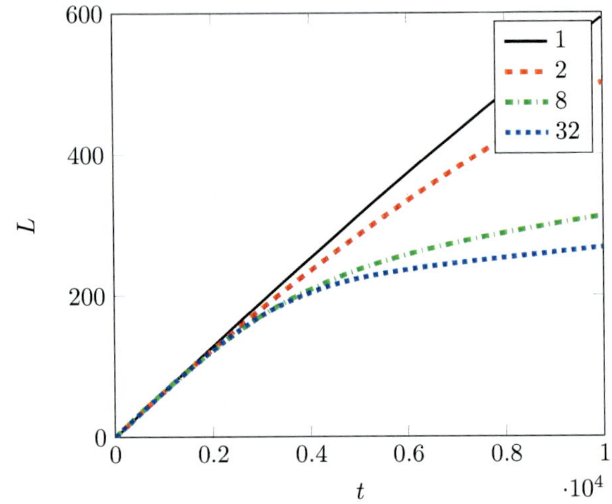

Fig. 7.8 Cumulative regret of weighted Bootstrap Q-learning for various values of bootstrap replicates (1 is equivalent to plain Q-learning). Generally speaking, an increased amount of replicates leads to improved exploration performance

1. Develop and test a simulation-based algorithm (such as Sarsa) for estimating Φ^{π}_{μ}, and prove its asymptotic convergence. *Hint: Focus on the fact that you'd like to estimate a value function for all possible reward functions.*
2. Consider a model-based approach building an empirical transition kernel $P^{\pi}_{\hat{\mu}}$. How good are your value function estimates in the first versus the second approach? Why would you expect either one to be better?
3. Can the same idea be extended to Q-learning?

References

1. Robbins, H., Monro, S.: A stochastic approximation method. Ann. Math. Stat. **22**(3), 400–407 (1951)
2. Bertsekas, D.P., Tsitsiklis, J.N.: Neuro-Dynamic Programming. Athena Scientific (1996)
3. Dearden, R., Friedman, N., Russell, S.J.: Bayesian Q-learning. In: Proceedings of the Fifteenth National Conference on Artificial Intelligence and Tenth Innovative Applications of Artificial Intelligence Conference, AAAI 98, IAAI98, pp. 761–768. AAAI Press, The MIT Press (1998)
4. Mnih, V., Kavukcuoglu, K., Silver, D., et al.: Human-level control through deep reinforcement learning. Nature **518**(7540), 529–533 (2015)
5. Tsitsiklis, J.N.: Asynchronous stochastic approximation and Q-learning. Mach. Learn. **16**(3), 185–202 (1994)
6. Watkins, C.J.C.H, Dayan, P: Q-learning. Mach. Learn. **8**(3–4), 279–292 (1992)
7. Kearns, M., Singh, S.: Finite sample convergence rates for Q-learning and indirect algorithms. In: Advances in Neural Information Processing Systems, vol. 11, pp. 996–1002. The MIT Press (1999)
8. Strehl, A.L., Li, L., Wiewiora, E., Langford, J., Littman, M.L.: Pac model-free reinforcement learning. In: Machine Learning, Proceedings of the 23rd International Conference, ICML 2006, pp. 881–888. ACM (2006)

9. Dimitrakakis, C.: Nearly optimal exploration-exploitation decision thresholds. In: Artificial Neural Networks–ICANN 2006, 16th International Conference, Proceedings, Part I
10. Dimitrakakis, C.: Ensembles for Sequence Learning. Ph.D. thesis, École Polytechnique Fédérale de Lausanne (2006)
11. Osband, I., Blundell, C., Pritzel, A., Van Roy, B.: Deep exploration via bootstrapped DQN. In: Advances in Neural Information Processing Systems, vol. 29, pp. 4026–4034 (2016)

Chapter 8
Approximate Representations

8.1 Introduction

In this chapter, we consider methods for approximating value functions, policies, or transition kernel. This is in particular useful when the state or policy space are large, so that one has to use some parametrization that may not include the true value function, policy, or transition kernel. In general, we shall assume the existence of either some approximate value function space $\mathcal{V}_\Theta$ or some approximate policy space Π_Θ, which are the set of allowed value functions and policies, respectively. For the purposes of this chapter, we will assume that we have access to some simulator or approximate model of the transition probabilities, wherever necessary. Model-based reinforcement learning where the transition probabilities are explicitly estimated will be examined in the next two chapters.

As an introduction, let us start with the case where we have a value function space $\mathcal{V}$ and some value function $\boldsymbol{v} \in \mathcal{V}$ that is our best approximation of the optimal value function. Then we can define the greedy policy with respect to $\boldsymbol{v}$ as follows:

Definition 8.1.1 (*$\boldsymbol{v}$-greedy policy and value function*)

$$\pi_{\boldsymbol{u}}^* \in \arg\max_{\pi \in \Pi} \mathscr{L}_\pi \boldsymbol{u}, \qquad \boldsymbol{v}_{\boldsymbol{u}}^* = \mathscr{L} \boldsymbol{u},$$

where $\pi : \mathcal{S} \to \Delta\mathcal{A}$ maps from states to action distributions.

Although the greedy policies need not be stochastic, here we are explicitly considering stochastic policies, because this sometimes facilitates finding a good approximation. If $\boldsymbol{u}$ is the optimal value function V^*, then the greedy policy is going to be optimal.

More generally, when we are trying to approximate a value function, we usually are constrained to look for it in a parametrized set of value functions $\mathcal{V}_\Theta$, where Θ is the parameter space. Hence, it might be the case that the optimal value function may not lie within $\mathcal{V}_\Theta$. Similarly, the policies that we can use lie in a space Π_Θ, which

C. Dimitrakakis and R. Ortner, *Decision Making Under Uncertainty and Reinforcement Learning*, Intelligent Systems Reference Library 223,
https://doi.org/10.1007/978-3-031-07614-5_8

may not include the greedy policy itself. This is usually because it is not possible to represent all possible value functions and policies in complex problems.

Usually, we are not aiming at a *uniformly good* approximation to a value function or policy. Instead, we define ϕ, a distribution on $\mathcal{S}$, which specifies on which parts of the state space we want to have a good approximation by placing higher weight on the most important states. Frequently, ϕ only has a *finite* support, meaning that we only measure the approximation error over a finite set of representative states $\hat{\mathcal{S}} \subseteq \mathcal{S}$. In the sequel, we shall always define the quality of an approximate value or policy with respect to ϕ.

In the remainder of this chapter, we shall examine a number of approximate dynamic programming algorithms. What all of these algorithms have in common is the requirement to calculate an approximate value function or policy. The two next sections given an overview of the basic problem of fitting an approximate value function or policy to a target.

8.1.1 Fitting a Value Function

Let us begin by considering the problem of finding the value function $\boldsymbol{v}_\theta \in \mathcal{V}_\Theta$ that best matches a target value function $\boldsymbol{u}$ that is not necessarily in $\mathcal{V}_\Theta$. This can be done by minimizing the difference between the target value $\boldsymbol{u}$ and the approximation v_θ, that is,

$$\|\boldsymbol{v}_\theta - \boldsymbol{u}\|_\phi = \int_{\mathcal{S}} |\boldsymbol{v}_\theta(s) - \boldsymbol{u}(s)| \, \mathrm{d}\phi(s)$$

with respect to some measure ϕ on $\mathcal{S}$. If $\boldsymbol{u} = V^*$, i.e., the optimal value function, then we end up getting the best possible value function with respect to the distribution ϕ. We can formalize the idea for fitting an approximate value function to a target as follows.

Approximate value function problem

Given $\mathcal{V}_\Theta = \{\boldsymbol{v}_\theta \mid \theta \in \Theta\}$,

$$\text{find } \theta^* \in \arg\min_{\theta \in \Theta} \|\boldsymbol{v}_\theta - \boldsymbol{u}\|_\phi,$$

where $\|\cdot\|_\phi \triangleq \int_{\mathcal{S}} |\cdot| \, \mathrm{d}\phi$.

Unfortunately, this minimization problem can be difficult to solve in general. A particularly simple case is when the set of approximate functions is small enough for the minimization to be performed via enumeration.

Example 8.1 (Fitting a finite number of value functions) Consider a finite space of value functions $\mathcal{V} = \{\boldsymbol{v}_1, \boldsymbol{v}_2, \boldsymbol{v}_3\}$, which we wish to fit to a target value function $\boldsymbol{u}$.

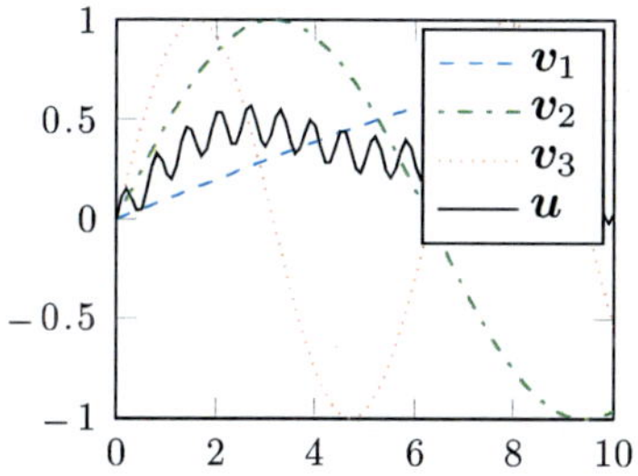

(a) The target function and the three candidates.

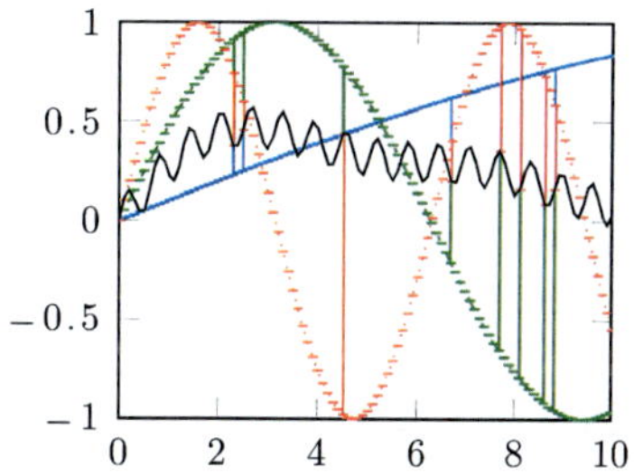

(b) The errors at the chosen points.

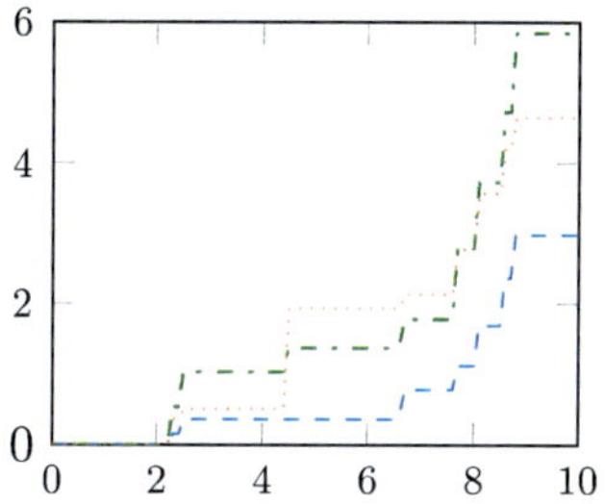

(c) The total error of each candidate.

Fig. 8.1 Fitting a value function in $\mathcal{V} = \{\boldsymbol{v}_1, \boldsymbol{v}_2, \boldsymbol{v}_3\}$ to a target value function $\boldsymbol{u}$, over a finite number of points. While none of the three candidates is a perfect fit, we clearly see that $\boldsymbol{v}_1$ has the lowest cumulative error over the measured set of points

In this particular scenario, $\boldsymbol{v}_1(x) = \sin(0.1x)$, $\boldsymbol{v}_2(x) = \sin(0.5x)$, $\boldsymbol{v}_3(x) = \sin(x)$, while

$$\boldsymbol{u}(x) = 0.5\sin(0.1x) + 0.3\sin(0.1x) + 0.1\sin(x) + 0.1\sin(10x).$$

Clearly, none of the given functions is a perfect fit. In addition, finding the best overall fit requires minimizing an integral. So, for this problem we choose a random set of points $X = \{x_t\}$ on which to evaluate the fit, with $\phi(x_t) = 1$ for every point $x_t \in X$. This is illustrated in Fig. 8.1, which shows the error of the functions at the selected points, as well as their cumulative error.

In the example above, the approximation space $\mathcal{V}$ does not have a member that is sufficiently close to the target value function. It could be that a larger function space contains a better approximation. However, it may be difficult to find the best fit in an arbitrary set $\mathcal{V}$.

8.1.2 Fitting a Policy

The problem of fitting a policy is not significantly different from that of fitting a value function, especially when the action space is continuous. Once more, we define an appropriate normed vector space so that it makes sense to talk about the normed difference between two policies π, π' with respect to some measure ϕ on the states, more precisely defined as

$$\left\|\pi - \pi'\right\|_\phi = \int_{\mathcal{S}} \|\pi(\cdot \mid s) - \pi'(\cdot \mid s)\| \, \mathrm{d}\phi(s),$$

where the norm within the integral is usually the L_1 norm. For a finite action space, this corresponds to $\|\pi(\cdot \mid s) - \pi'(\cdot \mid s)\| = \sum_{a \in \mathcal{A}} |\pi(a \mid s) - \pi'(a \mid s)|$, but certainly other norms may be used and are sometimes more convenient. The optimization problem corresponding to fitting an approximate policy from a set of policies Π_Θ to a target policy π is shown below.

The policy approximation problem

Given $\Pi_\Theta = \{\pi_\theta \mid \theta \in \Theta\}$,

$$\text{find } \theta^* \in \arg\min_{\theta \in \Theta} \|\pi_\theta - \pi^*_{\boldsymbol{u}}\|_\phi,$$

where $\pi^*_{\boldsymbol{u}} = \arg\max_{\pi \in \Pi} \mathscr{L}_1 \boldsymbol{u}$.

Once more, the minimization problem may not be trivial, but there are some cases where it is particularly easy. One of these is when the policies can be efficiently enumerated, as in the example below.

Example 8.2 (Fitting a finite space of policies) For simplicity, consider the space of deterministic policies with a binary action space $\mathcal{A} = \{0, 1\}$. Then each policy can be represented as a simple mapping $\pi : \mathcal{S} \to \{0, 1\}$, corresponding to a binary partition of the state space. In this example, the state space is the 2-dimensional unit cube, $\mathcal{S} = [0, 1]^2$. Figure 8.2 shows an example policy, where the light red and light green areas represent taking action 1 and 0, respectively. The measure ϕ has support only on the crosses and circles, which indicate the action taken at that location. Consider a policy space Π consisting of just four policies. Each set of two policies is indicated by the magenta (dashed) and blue (dotted) lines in Fig. 8.2. Each line corresponds to two possible policies, one selecting action 1 in the high region, and the other selecting action 0 instead. In terms of our error metric, the best policy is the one that makes the fewest mistakes. Consequently, the best policy in this set to use the blue line and play action 1 (red) in the top-right region.

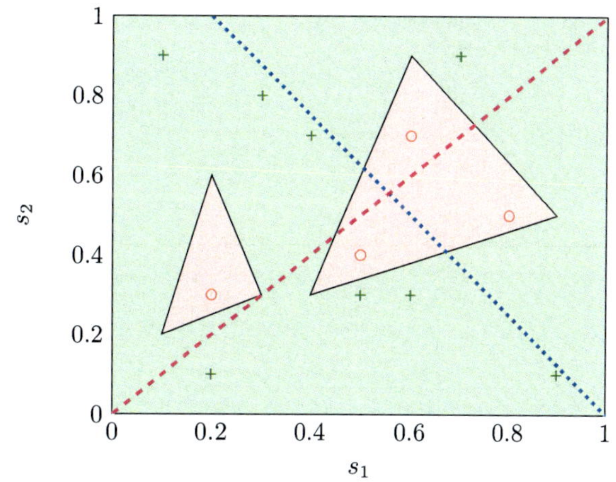

Fig. 8.2 An example policy. The red areas indicate taking action 1, and the green areas action 0. The ϕ measure has finite support, indicated by the crosses and circles. The blue and magenta lines indicate two possible policies that separate the state space with a hyperplane

8.1.3 Features

Frequently, when dealing with large, or complicated spaces, it pays off to project (observed) states and actions onto a feature space $\mathcal{X}$. In that way, we can make problems much more manageable. Generally speaking, a feature mapping is defined as follows.

Feature mapping

For $\mathcal{X} \subset \mathbb{R}^n$, a feature mapping $f : \mathcal{S} \times \mathcal{A} \to \mathcal{X}$ can be written in vector form as

$$f(s, a) = \begin{bmatrix} f_1(s, a) \\ \dots \\ f_n(s, a) \end{bmatrix}.$$

Obviously, one can define feature mappings $f : \mathcal{S} \to \mathcal{X}$ for states only in a similar manner.

What sort of functions should we use? A common idea is to use a set of smooth symmetric functions, such as usual radial basis functions.

Example 8.3 (Radial Basis Functions) Let d be a metric on $\mathcal{S} \times \mathcal{A}$ and define the a set of characteristic state-action pairs $\{(s_i, a_i) \mid i = 1, \dots, n\}$. These can act as centers for a set of radial basis functions, defined as follows:

$$f_i(s, a) \triangleq \exp\{-d[(s, a), (s_i, a_i)]\}$$

These functions are sometimes called *kernels*. A one-dimensional example of Gaussian radial basis functions is shown in Fig. 8.3.

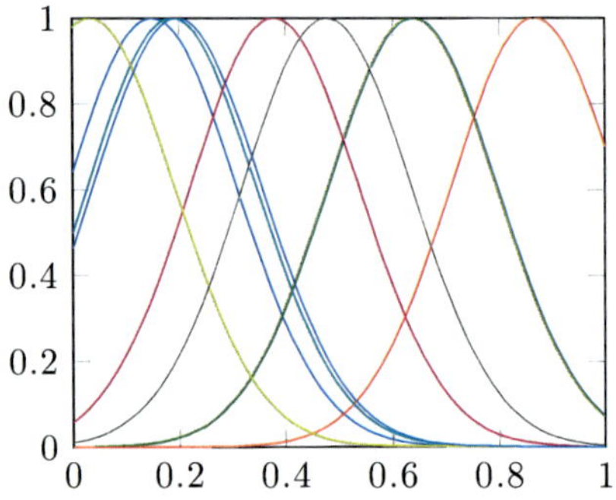

Fig. 8.3 Radial basis functions

Another common type of functions are binary functions. These effectively discretize a continuous space through either a cover or a partition.

Definition 8.1.2 A collection of sets $\mathcal{G}$ is a *cover* of X iff $\bigcup_{S\in\mathcal{G}} S \supset X$.

Definition 8.1.3 A collection of sets $\mathcal{G}$ is a *partition* of X iff

1. If $S \neq R \in \mathcal{G}$, then $S \cap R = \emptyset$.
2. $\bigcup_{S\in\mathcal{G}} S = X$.

In reinforcement learning, feature functions corresponding to partitions are usually referred to as tilings.

Example 8.4 (Tilings) Let $\mathcal{G} = \{X_1, \ldots, X_n\}$ be a *partition* of $\mathcal{S} \times \mathcal{A}$. Then the f_i can be defined by

$$f_i(s, a) \triangleq \mathbb{I}\{(s, a) \in X_i\}.$$

Multiple tilings create a cover and can be used without many difficulties with most discrete reinforcement learning algorithms, cf. Sutton and Barto [1].

8.1.4 Estimation Building Blocks

Now that we have looked at the basic problems in approximate regimes, let us look at some methods for obtaining useful approximations. First of all, we introduce some basic concepts such as look-ahead and rollout policies for estimating value functions. Then we formulate value function approximation and policy estimation as an optimization problem. These are going to be used in the remaining sections. For example, Sect. 8.2 introduces the well known approximate policy iteration algorithm, which combines those two steps into approximate policy evaluation and approximate policy improvement.

8.1.4.1 Look-ahead Policies

Given an approximate value function $\boldsymbol{u}$, the transition model P_μ of the MDP and the expected rewards r_μ, we can always find the improving policy given in Definition 8.1.1 via the following single-step look-ahead.

Single-step look-ahead

Let $q(i, a) \triangleq r_\mu(i, a) + \gamma \sum_{j \in \mathcal{S}} P_\mu(j \mid i, a)\, \boldsymbol{u}(j)$. Then the single-step look-ahead policy is defined as

$$\pi_{\boldsymbol{q}}(a \mid i) > 0 \qquad \text{iff } a \in \arg\max_{a' \in \mathcal{A}} q(i, a').$$

(Note that there may be more than one maximizing action.)

We are however not necessarily limited to the first-step. By looking T steps forward into the future we can improve both our value function and policy estimates.

T-step look-ahead

Define $\boldsymbol{u}_k$ recursively as:

$$\begin{aligned}
\boldsymbol{u}_0 &= \boldsymbol{u}, \\
q_k(i, a) &= r_\mu(i, a) + \gamma \sum_{j \in \mathcal{S}} P_\mu(j \mid i, a)\, \boldsymbol{u}_{k-1}(j), \\
\boldsymbol{u}_k(i) &= \max \{q_k(i, a) \mid a \in \mathcal{A}\}.
\end{aligned}$$

Then the T-step look-ahead policy is defined by

$$\pi_{\boldsymbol{q}_t}(a \mid i) > 0 \qquad \text{iff } a \in \arg\max_{a' \in \mathcal{A}} \boldsymbol{q}_T(i, a').$$

In fact, taking $\boldsymbol{u} = \mathbf{0}$, this recursion is identical to solving the k-horizon problem and in the limit we obtain solution to the original problem. In the general case, our value function estimation error is bounded by $\gamma^k \|\boldsymbol{u} - V^*\|$.

8.1.4.2 Rollout Policies

As we have seen in Sect. 7.2.2 one way to obtain an approximate value function of an arbitrary policy π is to use Monte Carlo estimation, that is, to simulate several sequences of state-action-reward tuples by running the policy on the MDP. More specifically, we have the following rollout estimate.

Rollout estimate of the q-factor

In particular, from each state i, we take K_i rollouts to estimate:

$$q(i, a) = \frac{1}{K_i} \sum_{k=1}^{K_i} \sum_{t=0}^{T_k - 1} r(s_{t,k}, a_{t,k}), \tag{8.1.1}$$

where $s_{t,k}, a_{t,k} \sim \mathbb{P}^{\pi}_{\mu}(\cdot \mid s_0 = i, a_0 = a)$, and $T_k \sim \mathcal{G}\rceil\wr\Updownarrow(1 - \gamma)$.

This results in a set of samples of $\boldsymbol{q}$-factors. The next problem is to find a parametric policy π_θ that approximates the greedy policy with respect to our samples, $\pi^*_{\boldsymbol{q}}$. For a finite number of actions, this can be seen as a classification problem [2]. For continuous actions, it becomes a regression problem. As indicated before we define a distribution ϕ on a set of *representative states* $\hat{S}$ over which we wish to perform the minimization.

Rollout policy estimation

Given some distribution ϕ on a set $\hat{S}$ of representative states and a set of samples $q(i, a)$, giving us the greedy policy

$$\pi^*_{\boldsymbol{q}}(a \mid i) = \arg\max q(i \mid a)$$

and a parametrized policy space $\{\pi_\theta \mid \theta \in \Theta\}$, the goal is to determine

$$\min_\theta \left\| \pi_\theta - \pi^*_{\boldsymbol{q}} \right\|_\phi .$$

8.1.5 The Value Estimation Step

We can now attempt to fit a parametric approximation to a given state or state-action value function. This is often better than simply maintaining a set of rollout estimates from individual states (or state-action pairs), as it might enable us to generalize over the complete state space. A simple parametrization for the value function is to use a generalized linear model on a set of features. Then the value function is a linear function of the features with parameters $\boldsymbol{\theta}$. More precisely, we can define the following model for the case where we have a feature mapping on states.

Generalized linear model using state features (or kernel)

Given a feature mapping $f : \mathcal{S} \to \mathbb{R}^n$ and parameters $\boldsymbol{\theta} \in \mathbb{R}^n$, compute the approximation

$$\boldsymbol{v}_{\boldsymbol{\theta}}(s) = \sum_{i=1}^{n} \theta_i f_i(s).$$

Choosing the model representation is only the first step. We now have to use it to represent a specific value function. In order to do this, as before we first pick a set of representative states $\hat{\mathcal{S}}$ to fit our value function $\boldsymbol{v}_{\boldsymbol{\theta}}$ to $\boldsymbol{v}$. This type of estimation can be seen as a regression problem, where the observations are value function measurements at different states.

Fitting a value function to a target

Let ϕ be a distribution over representative states $\hat{\mathcal{S}}$. For some constants κ, $p > 0$, we define the weighted prediction error per state as

$$c_s(\boldsymbol{\theta}) = \phi(s) \left\| \boldsymbol{v}_{\boldsymbol{\theta}}(s) - \boldsymbol{u}(s) \right\|_p^{\kappa},$$

where the total prediction error is $c(\boldsymbol{\theta}) = \sum_{s \in \hat{\mathcal{S}}} c_s(\boldsymbol{\theta})$.
The goal is to find a $\boldsymbol{\theta}$ minimizing $c(\boldsymbol{\theta})$.

Minimizing this error can be done using gradient descent, which is a general algorithm for finding local minima of smooth cost functions. Generally, minimizing a real-valued cost function $c(\boldsymbol{\theta})$ with gradient descent involves an algorithm iteratively approximating the value minimizing c:

$$\boldsymbol{\theta}^{(n+1)} = \boldsymbol{\theta}^{(n)} + \alpha_n \nabla c(\boldsymbol{\theta}^{(n)})$$

Under certain conditions[1] on the step-size parameter α_n, $\lim_{n \to \infty} c(\boldsymbol{\theta}^{(n)}) = \min_{\boldsymbol{\theta}} c(\boldsymbol{\theta})$.

Example 8.5 (Gradient descent for $p = 2, \kappa = 2$) For $p = 2, \kappa = 2$ the square root and κ cancel out and we obtain

$$\nabla_{\boldsymbol{\theta}} c_s = \phi(s) \sum_{a \in \mathcal{A}} \nabla_{\boldsymbol{\theta}} [\boldsymbol{v}_{\boldsymbol{\theta}}(s) - \boldsymbol{u}(s)]^2 = 2[\boldsymbol{v}_{\boldsymbol{\theta}}(s) - \boldsymbol{u}(s)] \nabla_{\boldsymbol{\theta}} \boldsymbol{v}_{\boldsymbol{\theta}},$$

where $\nabla_{\boldsymbol{\theta}} \boldsymbol{v}_{\boldsymbol{\theta}}(s) = f(s)$. Taking partial derivatives $\partial / \partial \theta_j$ leads to the update rule

[1] See also Sect. 7.1.1.

$$\theta_j^{(n+1)} = \theta_j^{(n)} - 2\alpha\phi(s)[\boldsymbol{v}_{\boldsymbol{\theta}^{(n)}}(s) - \boldsymbol{u}(s)]f_j(s).$$

However, the value function is not necessarily self-consistent, meaning that we do not have the identity $\boldsymbol{v}_\theta(s) = \boldsymbol{r}(s) + \int_S \boldsymbol{v}_\theta(s') \, \mathrm{d}P(s' \mid s, a)$. For that reason, we can instead choose a parameter that tries to make the parametrized value function self-consistent by minimizing the Bellman error.

Minimizing the Bellman error

For a given ϕ and $\hat{P}$, the goal is to find $\boldsymbol{\theta}$ minimizing

$$b(\theta) \triangleq \left\| r(s, a) + \gamma \int_S \boldsymbol{v}_\theta(s') \, \mathrm{d}\hat{P}(s' \mid s, a) - \boldsymbol{v}_\theta(s) \right\|_\phi . \tag{8.1.2}$$

Here $\hat{P}$ is not necessarily the true transition kernel. It can be a model or an empirical approximation (in which case the integral would only be over the empirical support). The summation itself is performed with respect to the measure ϕ.

In this chapter, we will look at two methods for approximately minimizing the Bellman error. The first, least square policy iteration is a batch algorithm for approximate policy iteration and finds the least-squares solution to the problem using the empirical transition kernel. The second is a gradient based method, which is flexible enough to use either an explicit model of the MDP or the empirical transition kernel.

It is also possible to simultaneously approximate some value function $\boldsymbol{u}$ and minimizing the Bellman error by considering the minimization problem

$$\min_{\boldsymbol{\theta}} c(\boldsymbol{\theta}) + \lambda b(\theta)$$

whereby the Bellman error acts as a regularizer ensuring that our approximation is indeed as consistent as possible.

8.1.6 Policy Estimation

A natural parametrization for policies is to use a generalized linear model on a set of features. Then a policy can be described as a linear function of the features with parameters $\boldsymbol{\theta}$, together with an appropriate link function. More precisely, we can define the following model.

Generalized linear model using features (or kernel)

Given a feature mapping $f : \mathcal{S} \times \mathcal{A} \to \mathbb{R}^n$, parameters $\boldsymbol{\theta} \in \mathbb{R}^n$, and a link function $\ell : \mathcal{S} \times \mathcal{A} \to \mathbb{R}^+$, the parametrized policy $\pi_{\boldsymbol{\theta}}$ is defined as

$$\pi_{\boldsymbol{\theta}}(a \mid s) = \frac{g_{\boldsymbol{\theta}}(s, a)}{h_{\boldsymbol{\theta}}(s)},$$

$$\text{where } g_{\boldsymbol{\theta}}(s, a) = \ell\left(\sum_{i=1}^{n} \theta_i f_i(s, a)\right) \text{ and } h_{\boldsymbol{\theta}}(s) = \sum_{b \in \mathcal{A}} g_{\boldsymbol{\theta}}(s, b).$$

The link function ℓ ensures that the denominator is positive, and the policy is a distribution over actions. An alternative method would be to directly constrain the policy parameters so the result is always a distribution, but this would result in a constrained optimization problem. A typical choice for the link function is $\ell(x) = \exp(x)$, which results in the softmax family of policies.

In order to fit a policy, we first pick a set of representative states $\hat{\mathcal{S}}$ and then find a policy $\pi_{\boldsymbol{\theta}}$ that approximates a target policy π, which is typically the greedy policy with respect to some value function. In order to do so, we can define an appropriate cost function and then estimate the optimal parameters via some arbitrary optimization method.

Fitting a policy through a cost function

Given a target policy π and a cost function

$$c_s(\boldsymbol{\theta}) = \phi(s) \left\| \pi_{\boldsymbol{\theta}}(\cdot \mid s) - \pi(\cdot \mid s) \right\|_p^{\kappa},$$

the goal is to find the parameter $\boldsymbol{\theta}$ minimizing $c(\boldsymbol{\theta}) = \sum_{s \in \hat{\mathcal{S}}} c_s(\boldsymbol{\theta})$.

Once more, we can use gradient descent to minimize the cost function. We obtain different results for different norms, but the three cases of main interest are $p = 1$, $p = 2$, and $p \to \infty$. We present the first one here, and leave the others as an exercise.

Example 8.6 (The case $p = 1$, $\kappa = 1$) The derivative can be written as

$$\nabla_{\boldsymbol{\theta}} c_s = \phi(s) \sum_{a \in \mathcal{A}} \nabla_{\boldsymbol{\theta}} |\pi_{\boldsymbol{\theta}}(a \mid s) - \pi(a \mid s)|,$$

$$\nabla_{\boldsymbol{\theta}} |\pi_{\boldsymbol{\theta}}(a \mid s) - \pi(a \mid s)| = \nabla_{\boldsymbol{\theta}} \pi_{\boldsymbol{\theta}}(a \mid s) \operatorname{sgn}[\pi_{\boldsymbol{\theta}}(a \mid s) - \pi(a \mid s)].$$

The policy derivative in turn is

$$\pi_\theta(a \mid s) = \frac{h(s)\nabla_\theta g(s,a) - \nabla_\theta h(s) g(s,a)}{h(s)^2},$$

with $\nabla_\theta h(s) = \left(\sum_{b\in\mathcal{A}} f_i(s,b)\right)_i$ and $\nabla_\theta g(s,a) = f(s,a)$. Taking partial derivatives $\partial/\partial\theta_j$, leads to the update rule

$$\theta_j^{(n+1)} = \theta_j^{(n)} - \alpha_n \phi(s) \left(\pi_{\theta^{(n)}}(a \mid s) \sum_{b\in\mathcal{A}} f_j(s,b) - f_j(s,a) \right).$$

Alternative cost functions. It is often a good idea to add a *penalty term* of the form $\|\boldsymbol{\theta}\|^q$ to the cost function, constraining the parameters to be small. The purpose of this is to prevent overfitting of the parameters for a small number of observations.

8.2 Approximate Policy Iteration (API)

The main idea of approximate policy iteration is to replace the exact Bellman operator $\mathscr{L}$ with an approximate version $\hat{\mathscr{L}}$ to obtain an approximate optimal policy and a respective approximate optimal value.

Just as in standard policy iteration introduced in Sect. 6.5.4.2, there is a policy improvement step and a policy evaluation step. In the policy improvement step, we simply try to get as close as possible to the best possible improvement using a restricted set of policies and an approximation of the Bellman operator. Similarly, in the policy evaluation step, we try to get as close as possible to the actual value of the improved policy using a respective set of value functions.

Algorithm 8.1 Generic approximate policy iteration algorithm

Input: initial value function approximation $\boldsymbol{v}_0$, approximate Bellman operator $\hat{\mathscr{L}}$, approximate value estimator $\hat{V}$, policy space $\hat{\Pi}$, value function space $\hat{\mathcal{V}}$, norms $\|\cdot\|_\phi$, and $\|\cdot\|_\psi$,
for $k = 1, \dots$ **do**
  $\pi_k = \arg\min_{\pi\in\hat{\Pi}} \left\| \hat{\mathscr{L}}_\pi \boldsymbol{v}_{k-1} - \mathscr{L}\boldsymbol{v}_{k-1} \right\|_\phi$ `// policy improvement`
  $\boldsymbol{v}_k = \arg\min_{\boldsymbol{v}\in\hat{\mathcal{V}}} \|\boldsymbol{v} - \hat{V}_\mu^{\pi_k}\|_\psi$ `// policy evaluation`
end for

At the k-th iteration of the policy improvement step the approximate value $\boldsymbol{v}_{k-1}$ of the previous policy π_{k-1} is used to obtain an improved policy π_k. However, note that we may not be able to implement the policy $\arg\max_\pi \mathscr{L}_\pi \boldsymbol{v}_{k-1}$ for two reasons. Firstly, the policy space $\hat{\Pi}$ may not include all possible policies. Secondly, the Bellman operator is in general also only an approximation. In the policy evaluation step, we aim at finding the function $\boldsymbol{v}_k$ that is the closest to the true value function of policy π_k. However, even if the value function space $\hat{\mathcal{V}}$ is rich enough, the minimization is done over a norm that integrates over a finite subset of the state space. The following section discusses the effect of those errors on the convergence of approximate policy iteration.

8.2.1 Error Bounds for Approximate Value Functions

If the approximate value function $\boldsymbol{u}$ is close to V^* then the greedy policy with respect to $\boldsymbol{u}$ is close to optimal. For a finite state and action space, the following holds.

Theorem 8.2.1 *Consider a finite MDP μ with discount factor $\gamma < 1$ and a vector $\boldsymbol{u} \in \mathcal{V}$ such that $\left\| \boldsymbol{u} - V_\mu^* \right\|_\infty = \epsilon$. If π is the $\boldsymbol{u}$-greedy policy then*

$$\left\| V_\mu^\pi - V_\mu^* \right\|_\infty \leq \frac{2\gamma\epsilon}{1-\gamma}.$$

In addition, $\exists \epsilon_0 > 0$ s.t. if $\epsilon < \epsilon_0$, then π is optimal.

Proof Recall that $\mathscr{L}$ is the one-step Bellman operator and $\mathscr{L}_\pi$ is the one-step policy operator on the value function. Then (skipping the index for μ)

$$\begin{aligned}
\left\| V^\pi - V^* \right\|_\infty &= \left\| \mathscr{L}_\pi V^\pi - V^* \right\|_\infty \\
&\leq \left\| \mathscr{L}_\pi V^\pi - \mathscr{L}_\pi \boldsymbol{u} \right\|_\infty + \left\| \mathscr{L}_\pi \boldsymbol{u} - V^* \right\|_\infty \\
&\leq \gamma \left\| V^\pi - \boldsymbol{u} \right\|_\infty + \left\| \mathscr{L} \boldsymbol{u} - V^* \right\|_\infty
\end{aligned}$$

by contraction, and by the fact that π is $\boldsymbol{u}$-greedy,

$$\begin{aligned}
&\leq \gamma \left\| V^\pi - V^* \right\|_\infty + \gamma \left\| V^* - \boldsymbol{u} \right\|_\infty + \gamma \left\| \boldsymbol{u} - V^* \right\|_\infty \\
&\leq \gamma \left\| V^\pi - V^* \right\|_\infty + 2\gamma\epsilon,
\end{aligned}$$

which proves the first part.

For the second part, note that the state and action sets are finite. Consequently, the set of policies is finite. Thus, there is some $\epsilon_0 > 0$ such that the best sub-optimal policy is ϵ_0-close to the optimal policy in value. So, if $\epsilon < \epsilon_0$, the obtained policy must be optimal. □

Building on this result, we can prove a simple bound for approximate policy iteration, assuming uniform error bounds on the approximation of the value of a policy as well as on the approximate Bellman operator. Even though these assumptions are quite strong, we still only obtain the following rather weak asymptotic convergence result.[2]

---

[2] For $\delta = 0$, this is identical to the result for ϵ-equivalent MDPs by Even-Dar and Mansour [3].

Theorem 8.2.2 (Bertsekas and Tsitsiklis [4], Proposition 6.2) *Assume that there are* ϵ, δ *such that, for all* k *the iterates* $\boldsymbol{v}_k, \pi_k$ *satisfy*

$$\|\boldsymbol{v}_k - V^{\pi_k}\|_\infty \le \epsilon,$$
$$\left\|\mathscr{L}_{\pi_{k+1}} \boldsymbol{v}_k - \mathscr{L} \boldsymbol{v}_k\right\|_\infty \le \delta.$$

Then

$$\limsup_{k\to\infty} \left\|V^{\pi_k} - V^*\right\|_\infty \le \frac{\delta + 2\gamma\epsilon}{(1-\gamma)^2}.$$

8.2.2 Rollout-Based Policy Iteration Methods

As suggested by Bertsekas and Tsitsiklis [4], one idea for estimating the value function is to simply perform rollouts, while the policy itself is estimated in parametric form. The first practical algorithm in this direction was Rollout Sampling Approximate Policy Iteration by Dimitrakakis and Lagoudakis [5]. The main idea is to concentrate rollouts in interesting parts of the state space, so as to maximize the expected amount of improvement we can obtain with a given rollout budget.

Algorithm 8.2 Rollout Sampling Approximate Policy Iteration

for $k = 1, \ldots$ **do**
  Select a set of representative states $\hat{S}_k$.
  for $n = 1, \ldots$ **do**
    Select a state $s_n \in \hat{S}_k$ maximizing $U_n(s)$ and perform a rollout obtaining $\{s_{t,k}, a_{t,k}\}$.
    If $\hat{a}^*(s_n)$ is optimal w.p. $1-\delta$, add s_n to $\hat{S}_k(\delta)$ and remove it from $\hat{S}_k$.
  end for
  Calculate $\boldsymbol{q}_k \approx Q^{\pi_k}$ from the rollouts, cf. Eq. (8.1.1).
  Train a classifier $\pi_{\theta_{k+1}}$ on the set of states $\hat{S}_k(\delta)$ with actions $\hat{a}^*(s)$.
end for

If we have data collects we can use the empirical state distribution to select starting states. In general, rollouts give us estimates $\boldsymbol{q}_k$, which are used to select states for further rollouts. That is, we compute for each state s actions

$$\hat{a}_s^* \triangleq \arg\max_a \boldsymbol{q}_k(s, a).$$

Then we select a state s_n maximizing the upper bound value $U_n(s)$ defined via

$$\hat{\Delta}_k(s) \triangleq \boldsymbol{q}_k(s, \hat{a}_s^*) - \max_{a \neq \hat{a}_s^*} \boldsymbol{q}_k(s, a),$$

$$U_n(s) \triangleq \hat{\Delta}_k - \max_{a \neq \hat{a}_s^*} \boldsymbol{q}_k(s, a) + \sqrt{\frac{1}{1 + c(s)}},$$

where $c(s)$ is the number of rollouts from state s. If the sampling of a state s stops whenever

$$\hat{\Delta}_k(s) \geq \sqrt{\frac{2}{c(s)(1-\gamma)^2} \ln\left(\frac{|\mathcal{A}| - 1}{\delta}\right)},$$

then we are certain that the optimal action has been identified with probability $1 - \delta$ for that state, due to Hoeffding's inequality. Unfortunately, guaranteeing a policy improvement for the complete state space is impossible, even with strong assumptions.[3]

8.2.3 *Least Squares Methods*

When considering quadratic error, it is tempting to use linear methods, such as least squares methods, which are very efficient. This requires to formulate the problem in linear form, using a feature mapping that projects individual states (or state-action pairs) onto a high-dimensional space. Then the value function can be represented as linear function of the parameters and this mapping, which minimizes a squared error over the observed trajectories.

To get an intuition for these methods, recall from Theorem 6.5.1 that the solution of

$$\boldsymbol{v} = \boldsymbol{r} + \gamma \boldsymbol{P}_{\mu,\pi} \boldsymbol{v}$$

is the value function of π and can be obtained via

$$\boldsymbol{v} = (\boldsymbol{I} - \gamma \boldsymbol{P}_{\mu,\pi})^{-1} \boldsymbol{r}.$$

Here we consider the setting where we do not have access to the transition matrix, but instead have some observations of transition (s_t, a_t, s_{t+1}). In addition, our state space can be continuous (e.g., $\mathcal{S} \subset \mathbb{R}^n$), so that the transition matrix becomes a general transition kernel. Consequently, the set of value functions $\mathcal{V}$ becomes a Hilbert space, while in the discrete setting a value function is simply a point in $\mathbb{R}^n$.

[3] First, note that if we need to identify the optimal action for k states, then the above stopping rule has an overall error probability of $k\delta$. In addition, even if we assume that value functions are smooth, it will be impossible to identify the boundary in the state space where the optimal policy should switch actions [6].

In general, we deal with this case via projections. We project from the infinite-dimensional Hilbert space to one with finite dimension on a subset of states: namely, the ones that we have observed. We also replace the transition kernel with the empirical transition matrix on the observed states.

Parametrization. Let us first deal with parametrizing a linear value function. Setting $\boldsymbol{v} = \boldsymbol{\Phi}\boldsymbol{\theta}$, where $\boldsymbol{\Phi}$ is a feature matrix and $\boldsymbol{\theta}$ is a parameter vector, we have

$$\boldsymbol{\Phi}\boldsymbol{\theta} = \boldsymbol{r} + \gamma \boldsymbol{P}_{\mu,\pi}\boldsymbol{\Phi}\boldsymbol{\theta},$$
$$\boldsymbol{\theta} = \left[(\boldsymbol{I} - \gamma \boldsymbol{P}_{\mu,\pi})\boldsymbol{\Phi}\right]^{-1} \boldsymbol{r}.$$

This simple linear parametrization of a value function is perfectly usable in a discrete MDP setting where the transition matrix is given. However, otherwise making use of this parametrization is not so straightforward. The first problem is how to define the transition matrix itself, since there is an infinite number of states. A simple solution to this problem is to only define the matrix on the observed sequence of states, and furthermore, so that the probability of transition to a particular state is 1 if that transition has been observed. This makes the matrix off-diagonal. More precisely, the construction is as follows.

Empirical construction. Given a set of data points $\{(s_t, a_t, r_t) \mid t = 1, \ldots, n\}$, we define:

1. the empirical reward vector $\boldsymbol{r} = (r_t)_t$,
2. the feature matrix $\boldsymbol{\Phi} = (\boldsymbol{\Phi}_t)_t$, with $\boldsymbol{\Phi}_t = f(s_t, a_t)$, and
3. the empirical $n \times n$ transition matrix $\boldsymbol{P}_{\mu,\pi}(s_t, s_{t'}) = \mathbb{I}\left\{t' = t + 1\right\}$.

Generally the value function space generated by the features and the linear parametrization does not allow us to obtain exact value functions. For this reason, instead of considering the inverse $\boldsymbol{A}^{-1}$ of the matrix $\boldsymbol{A} = (\boldsymbol{I} - \gamma \boldsymbol{P}_{\mu,\pi})\boldsymbol{\Phi}$ we use the *pseudo-inverse* defined as

$$\widetilde{\boldsymbol{A}}^{-1} \triangleq \boldsymbol{A}^{\top} \left(\boldsymbol{A}\boldsymbol{A}^{\top}\right)^{-1}.$$

If the inverse exists, then it is equal to the pseudo-inverse. However, in our setting, the matrix can be low rank, in which case we instead obtain the matrix minimizing the squared error, which in turn can be used to obtain a good estimate for the parameters. This immediately leads to the Least Squares Temporal Difference (LSTD) algorithm [7], which estimates an approximate value function for some policy π given some data D and a feature mapping f.

State-action value functions. As estimating a state-value function is not directly useful for obtaining an improved policy without a model, we can instead estimate a state-action value function as follows:

$$
\begin{aligned}
\boldsymbol{q} &= \boldsymbol{r} + \gamma \boldsymbol{P}_{\mu,\pi} \boldsymbol{q} \\
\boldsymbol{\Phi}\boldsymbol{\theta} &= \boldsymbol{r} + \gamma \boldsymbol{P}_{\mu,\pi} \boldsymbol{\Phi}\boldsymbol{\theta} \\
\boldsymbol{\theta} &= \widetilde{\big((\boldsymbol{I} - \gamma \boldsymbol{P}_{\mu,\pi})\boldsymbol{\Phi}\big)^{-1}} \boldsymbol{r}
\end{aligned}
$$

However, this approach has two drawbacks. The first is that it is difficult to get an unbiased estimate of θ. The second is that when we apply the Bellman operator to $\boldsymbol{q}$, the result may lie outside the space spanned by the features. For this reason, we instead consider the least-square projection $\boldsymbol{\Phi}(\boldsymbol{\Phi}^\top \boldsymbol{\Phi})^{-1}\boldsymbol{\Phi}^\top$, i.e.,

$$
\boldsymbol{q} = \boldsymbol{\Phi}(\boldsymbol{\Phi}^\top \boldsymbol{\Phi})^{-1}\boldsymbol{\Phi}^\top \left(\boldsymbol{r} + \gamma \boldsymbol{P}_{\mu,\pi} \boldsymbol{q}\right).
$$

Replacing $\boldsymbol{q} = \boldsymbol{\Phi}\boldsymbol{\theta}$ leads to the estimate

$$
\boldsymbol{\theta} = \left(\boldsymbol{\Phi}^\top (\boldsymbol{I} - \gamma \boldsymbol{P}_{\mu,\pi})\, \boldsymbol{\Phi}\right)^{-1} \boldsymbol{\Phi}^\top \boldsymbol{r}.
$$

In practice, of course, we do not have the transitions $\boldsymbol{P}_{\mu,\pi}$ but estimate them from data. Note that for any deterministic policy π and a set of T data points $(s_t, a_t, r_t, s'_t)_{t=1}^T$, we have

$$
\begin{aligned}
\boldsymbol{P}_{\mu,\pi}\boldsymbol{\Phi} &= \sum_{s'} P(s' \mid s, a)\, \boldsymbol{\Phi}(s', \pi(s')) \\
&\approx \frac{1}{T}\sum_{t=1}^{T} \hat{P}(s'_t \mid s_t, a_t)\, \boldsymbol{\Phi}(s'_t, \pi(s'_t)),
\end{aligned}
$$

where for $\hat{P}$ one can take the simple empirical transition matrix mentioned previously. This equation can be used to maintain $\boldsymbol{q}$-factors with $\boldsymbol{q}(s, a) = f(s, a)\boldsymbol{\theta}$ to obtain an empirical estimate of the Bellman operator as summarized in Algorithm 8.3.

Algorithm 8.3 LSTDQ-Least Squares Temporal Differences

input data $D = \{(s_t, a_t, r_t, s'_t) \mid t = 1, \dots, T\}$, feature mapping f, policy π
$\boldsymbol{A} = \sum_{t=1}^n \boldsymbol{\Phi}(s_t, a_t)\boldsymbol{P}[\boldsymbol{\Phi}(s_t, a_t) - \gamma\boldsymbol{\Phi}(s'_t, \pi(s'_t))]$
$\boldsymbol{b} = \sum_{t=1}^n \boldsymbol{\Phi}(s_t, a_t) r_t$
$\boldsymbol{\theta} = \boldsymbol{A}^{-1}\boldsymbol{b}$

The algorithm can be easily extended to approximate policy iteration, resulting in the well-known Least Squares Policy Iteration (LSPI) [8] algorithm shown in Algorithm 8.4. The idea is to repeatedly estimate the value function for improved policies using a least squares estimate, and then compute the greedy policy for each estimate.

Algorithm 8.4 LSPI-Least Squares Policy Iteration

input data $D = \{(s_t, a_t, r_t, s'_t) \mid t = 1, \ldots, T\}$, feature mapping f
Set π_0 arbitrarily.
for $k = 1, \ldots$ **do**
  $\boldsymbol{\theta}^{(k)} = \text{LSTDQ}(D, f, \pi_{k-1})$
  $\pi^{(k)} = \pi^*_{\boldsymbol{\Phi}\boldsymbol{\theta}^{(k)}}$
end for

8.3 Approximate Value Iteration

Approximate algorithms can also be defined for backwards induction. The general algorithmic structure remains the same as exact backwards induction, however the exact steps are replaced by approximations. Applying approximate value iteration may be necessary for two reasons. Firstly, it may not be possible to update the value function for all states. Secondly, the set of available value function representations may be not complex enough to capture the true value function.

8.3.1 Approximate Backwards Induction

The first algorithm is approximate backwards induction. Let us start with the basic backwards induction algorithm:

$$V_t^*(s) = \max_{a \in \mathcal{A}} \left\{ r(s, a) + \gamma \sum_{s'} V_{t+1}^*(s') \boldsymbol{P}_\mu(s'|s, a) \right\}$$

This is essentially the same both for finite and infinite-horizon problems. If we have to pick the value function from a set of functions $\mathcal{V}$, we can use the following value function approximation.

Let our estimate at time t be $\boldsymbol{v}_t \in \mathcal{V}$, with $\mathcal{V}$ being a set of (possibly parametrized) functions. Let $\hat{V}_t$ be our one-step update given the value function approximation at the next step, $\boldsymbol{v}_{t+1}$. Then $\boldsymbol{v}_t$ will be the closest approximation in that set.

Iterative approximation

$$\hat{V}_t(s) = \max_{a \in \mathcal{A}} \left\{ r(s, a) + \gamma \sum_{s'} \boldsymbol{P}_\mu(s' \mid s, a)\, \boldsymbol{v}_{t+1}(s') \right\}$$

$$\boldsymbol{v}_t = \arg\min_{\boldsymbol{v} \in \mathcal{V}} \left\| \boldsymbol{v} - \hat{V}_t \right\|$$

The above minimization can for example be performed by gradient descent. Consider the case where $\boldsymbol{v}$ is a parametrized function from a set of parametrized value functions $\mathcal{V}_\Theta$ with parameters $\boldsymbol{\theta}$. Then it is sufficient to maintain the parameters $\boldsymbol{\theta}^{(t)}$ at any time t. These can be updated with a gradient scheme at every step. In the online case, our next-step estimates can be obtained by gradient descent using a step size sequence $(\alpha_t)_t$.

Online gradient estimation

$$\boldsymbol{\theta}_{t+1} = \boldsymbol{\theta}_t - \alpha_t \nabla_{\boldsymbol{\theta}} \left\| \boldsymbol{v}_t - \hat{V}_t \right\|$$

This gradient descent algorithm can also be made stochastic, if we sample s' from the probability distribution $\boldsymbol{P}_\mu(s' \mid s, a)$ used in the iterative approximation. The next sections give some examples.

8.3.2 State Aggregation

In state aggregation, multiple different states with identical properties (with respect to rewards and transition probabilities) are identified in order to obtain a new aggregated state in an aggregated MDP with smaller state space. Unfortunately, it is very rarely the case that aggregated states are really indistinguishable with respect to rewards and transition probabilities. Nevertheless, as we can see in the example below, aggregation can significantly simplify computation through the reduction of the size of the state space.

Example 8.7 (Aggregated value function) A simple method for aggregation is to set the value of every state in an aggregate set to be the same. More precisely, let $\mathcal{G} = \{S_1, \ldots, S_n\}$ be a partition of $\mathcal{S}$, with $\boldsymbol{\theta} \in \mathbb{R}^n$ and let $f_k(s_t) = \mathbb{I}\{s_t \in S_k\}$. Then the approximate value function is

$$\boldsymbol{v}(s) = \boldsymbol{\theta}(k), \qquad \text{if } s \in S_k, k \neq 0,$$

where $\boldsymbol{\theta}(k)$ is the k th co-ordinate of $\boldsymbol{\theta}$.

In the above example, the value of every state corresponds to the value of the k-th set in the partition. Of course, this is only a very rough approximation if the sets S_k are very large. However, this is a convenient approach to use for gradient descent updates, as only one parameter needs to be updated at every step.

Online gradient estimate for aggregated value functions

Consider the case $\|\cdot\| = \|\cdot\|_2^2$. For $s_t \in S_k$ and some step size sequence $(\alpha_t)_t$:

$$\boldsymbol{\theta}_{t+1}(k) = (1 - \alpha_t)\boldsymbol{\theta}_t(k) + \alpha_t \max_{a \in \mathcal{A}} r(s_t, a) + \gamma \sum_{j \in \hat{S}} P(j \mid s_t, a)\, \boldsymbol{\theta}_t(f_k(j)),$$

while $\boldsymbol{\theta}_{t+1}(k) = \boldsymbol{\theta}(k)$ for $s_t \notin S_k$, where $\hat{S}$ is a subset of states.

Of course, whenever we perform the estimation online, we are limited to estimation on the sequence of states s_t that we visit. Consequently, estimation on other states may not be very good. It is indeed possible that we suffer from convergence problems as we alternate between estimating the values of different states in the aggregation. In the online algorithm, we posited the existence of a subset of states that we can use to perform the gradient update. We can formalize this through the notion of a representative state approximation.

8.3.3 *Representative State Approximation*

A more refined approach is to choose some representative states and try to approximate the value function of all other states as a convex combination of the value of the representative states.

Representative state approximation

Let $\hat{S}$ be a set of n representative states and $\boldsymbol{\theta} \in \mathbb{R}^n$ and a feature mapping f with

$$\sum_{i=1}^{n} f_i(s) = 1, \qquad \forall s \in \mathcal{S}.$$

The feature mapping is used to perform the convex combination. Usually, $f_i(s)$ is larger for representative states i which are "closer" to s. In general, the feature mapping is fixed, and we want to find a set of parameters for the values of the representative states. At time t, for each representative state i, we obtain a new estimate of its value function and plug it back in.

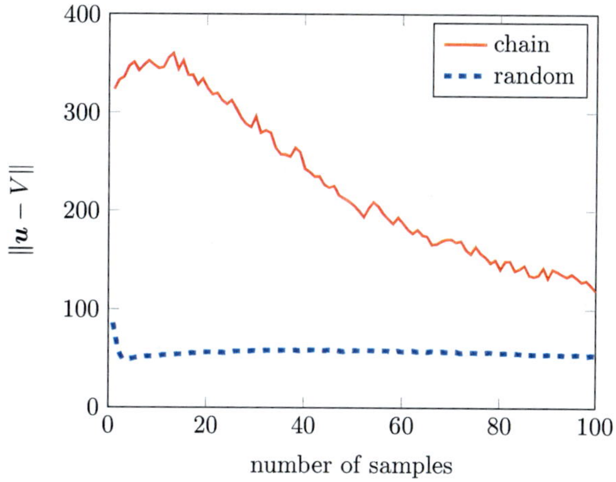

Fig. 8.4 Error in the representative state approximation for two different MDPs structures as we increase the number of sampled states. The first is the chain environment from Example 7.2, extended to 100 states. The second involves randomly generated MDPs with two actions and 100 states

> **Representative state update**
>
> For $i \in \hat{\mathcal{S}}$:
>
> $$\boldsymbol{\theta}_{t+1}(i) = \max_{a \in \mathcal{A}} \left\{ r(i, a) + \gamma \int \boldsymbol{v}_t(s) \, \mathrm{d}P(s \mid i, a) \right\} \tag{8.3.1}$$
>
> with
>
> $$\boldsymbol{v}_t(s) = \sum_{i=1}^{n} f_i(s) \boldsymbol{\theta}_t(i).$$

When the integration in (8.3.1) is not possible, we may instead approximate the expectation with a Monte Carlo method. One particular problem with this method arises when the transition kernel is very sparse. Then we are basing our estimates on approximate values of other states, which may be very far from any other representative state. This is illustrated in Fig. 8.4, which presents the value function error for the chain environment of Example 7.2 and random MDPs. Due to the linear structure of the chain environment, the states are far from each other. In contrast, the random MDPs are generally both quite dense and the state distribution for any particular policy mixes rather fast. Thus, states in the former tend to have very different values and in the latter very similar ones.

8.3.4 Bellman Error Methods

The problems with the representative state update can be alleviated through Bellman error minimization. The idea here is to obtain a value function that is as *consistent* as possible. The basic Bellman error minimization is given by

$$\min_{\theta} \|\boldsymbol{v}_{\theta} - \mathscr{L}\boldsymbol{v}_{\theta}\| .$$

This is different from the approximate backwards induction algorithm we saw previously, since the same parameter $\boldsymbol{\theta}$ appears in both terms inside the norm. Furthermore, if the norm has support in all of the state space and the approximate value function space contains the actual set of value functions then the minimum is 0 and we obtain the optimal value function.

Gradient update

For the L2-norm, we have the following α-step update:

$$\|\boldsymbol{v}_{\theta} - \mathscr{L}\boldsymbol{v}_{\theta}\| = \sum_{s \in \hat{S}} D_{\theta}(s)^2, \quad D_{\theta}(s) = \boldsymbol{v}_{\theta}(s) - \max_{a \in \mathcal{A}} \int_{S} \boldsymbol{v}_{\theta}(j)\, \mathrm{d}P(j \mid s, a)$$

Then the gradient update becomes $\boldsymbol{\theta}_{t+1} = \boldsymbol{\theta}_t - \alpha D_{\theta_t}(s_t) \nabla_{\theta} D_{\theta_t}(s_t)$, where

$$\nabla_{\theta} D_{\theta_t}(s_t) = \nabla_{\theta} \boldsymbol{v}_{\theta_t}(s_t) - \int_{S} \nabla_{\theta} \boldsymbol{v}_{\theta_t}(j)\, \mathrm{d}P(j \mid s_t, a_t^*)$$

with $a_t^* \triangleq \arg\max_{a \in \mathcal{A}} \left\{ r(s_t, a) + \gamma \int_S \boldsymbol{v}_{\theta}(j)\, \mathrm{d}P(j \mid s_t, a) \right\}$.

We can also construct a $\boldsymbol{q}$-factor approximation for the case where no model is available. This can be simply done by replacing P with the empirical transition observed at time t.

8.4 Policy Gradient

In the previous section, we have seen how to use gradient methods for value function approximation. It is also possible to use these methods to estimate policies—the only necessary ingredients are a policy representation and a way to evaluate a policy. The representation is usually parametric, but non-parametric representations are also possible. A common choice for parametrized policies is to use a feature function $f : \mathcal{S} \times \mathcal{A} \to \mathbb{R}^k$ and a linear parametrization with parameters $\theta \in \mathbb{R}^k$ leading to the following Softmax distribution:

$$\pi(a \mid s) = \frac{e^{F(s,a)}}{\sum_{a' \in \mathcal{A}} e^{F(s,a')}}, \qquad F(s, a) \triangleq \theta^{\top} f(s, a) \tag{8.4.1}$$

As usual, we would like to find a policy maximizing expected utility. Policy gradient algorithms employ gradient ascent on the expected utility to find a locally maximizing policy. Here we focus on the discounted reward criterion with discount factor γ, where a policy's expected utility is defined with respect to a starting state distribution $\boldsymbol{y}$ so that

$$\mathbb{E}^{\pi}_{\boldsymbol{y}}(U) = \sum_{s} y(s) V^{\pi}(s) = \sum_{s} y(s) \sum_{h} \mathbb{P}^{\pi}(h \mid s_1 = s)\, U(h), \tag{8.4.2}$$

where $U(h)$ is the utility of a trajectory h. This definition leads to a number of simple expressions for the gradient of the expected utility, including what is known as the policy gradient theorem of [9]. For notational simplicity, we omit the subscript $\boldsymbol{\theta}$ for the policy π.

Theorem 8.4.1 *Assuming that the reward only depends on the state, for any $\boldsymbol{\theta}$-parametrized policy space Π, the gradient of the utility from starting state distribution $\boldsymbol{y}$ can be equivalently written in the three following forms:*

$$\nabla_{\boldsymbol{\theta}} \mathbb{E}^{\pi}_{\boldsymbol{y}} U = \boldsymbol{y}^{\top} \gamma (\boldsymbol{I} - \gamma \boldsymbol{P}^{\pi}_{\mu})^{-1} \nabla_{\boldsymbol{\theta}} \boldsymbol{P}^{\pi}_{\mu} (\boldsymbol{I} - \gamma \boldsymbol{P}^{\pi}_{\mu})^{-1} \boldsymbol{r} \tag{8.4.3}$$

$$= \sum_{s} x^{\pi,\gamma}_{\mu,\boldsymbol{y}}(s) \sum_{a} \nabla_{\boldsymbol{\theta}} \pi(a \mid s)\, Q^{\pi}_{\mu}(s, a) \tag{8.4.4}$$

$$= \sum_{h} U(h)\, \mathbb{P}^{\pi}_{\mu}(h) \nabla_{\boldsymbol{\theta}} \ln \mathbb{P}^{\pi}_{\mu}(h), \tag{8.4.5}$$

where (as in Sect. 6.5.4.6) we use

$$x^{\pi,\gamma}_{\mu,\boldsymbol{y}}(s) = \sum_{s'} \sum_{t=0}^{\infty} \gamma^{t}\, \mathbb{P}^{\pi}_{\mu}(s_t = s \mid s_0 = s')\, y(s')$$

to denote the γ-discounted sum of state visits. Further, $h \in (\mathcal{S} \times \mathcal{A})^{*}$ denotes a state-action history, $\mathbb{P}^{\pi}_{\mu}(h)$ its probability under the policy π in the MDP μ, and $U(h)$ is the utility of history h.

Proof We begin by proving the claim (8.4.3). Note that

$$\mathbb{E}^{\pi}_{\mu} U = \boldsymbol{y}^{\top} (\boldsymbol{I} - \gamma \boldsymbol{P}^{\pi}_{\mu})^{-1} \boldsymbol{r},$$

where $\boldsymbol{y}$ is a starting state distribution vector and $\boldsymbol{P}^{\pi}_{\mu}$ is the transition matrix resulting from applying policy π to μ. Computing the derivative using matrix calculus gives

$$\nabla_{\boldsymbol{\theta}} \mathbb{E}\, U = \boldsymbol{y}^{\top} \nabla_{\boldsymbol{\theta}} (\boldsymbol{I} - \gamma \boldsymbol{P}^{\pi}_{\mu})^{-1} \boldsymbol{r},$$

as the only term involving $\boldsymbol{\theta}$ is π. The derivative of the matrix inverse can be written as

$$\nabla_\theta (\boldsymbol{I} - \gamma \boldsymbol{P}^\pi_\mu)^{-1} = -(\boldsymbol{I} - \gamma \boldsymbol{P}^\pi_\mu)^{-1} \nabla_\theta (\boldsymbol{I} - \gamma \boldsymbol{P}^\pi_\mu)(\boldsymbol{I} - \gamma \boldsymbol{P}^\pi_\mu)^{-1}$$
$$= \gamma (\boldsymbol{I} - \gamma \boldsymbol{P}^\pi_\mu)^{-1} \nabla_\theta \boldsymbol{P}^\pi_\mu (\boldsymbol{I} - \gamma \boldsymbol{P}^\pi_\mu)^{-1},$$

which concludes the proof of (8.4.3).

We proceed expanding $\boldsymbol{P}^\pi_\mu$ term, thus obtaining a formula that only has a derivative for π:

$$\frac{\partial}{\partial \theta_i} \mathbb{P}^\pi_\mu(s' \mid s) = \sum_a \mathbb{P}_\mu(s' \mid s, a) \frac{\partial}{\partial \theta_i} \pi(a \mid s)$$

Defining the state visitation matrix $\boldsymbol{X} \triangleq (\boldsymbol{I} - \gamma \boldsymbol{P})^{-1}$ we have, rewriting (8.4.3):

$$\nabla_\theta \mathbb{E}^\pi_\mu U = \gamma \boldsymbol{y}^\top \boldsymbol{X} \nabla_\theta \boldsymbol{P}^\pi_\mu \boldsymbol{X} \boldsymbol{r}$$

We are now ready to prove claim (8.4.4). Define the expected state visitation from the starting distribution to be $\boldsymbol{x} \triangleq \boldsymbol{y}^\top \boldsymbol{X}$, so that we obtain

$$\nabla_\theta \mathbb{E}^\pi_\mu U = \gamma \boldsymbol{x}^\top \nabla_\theta \boldsymbol{P}^\pi_\mu \boldsymbol{X} \boldsymbol{r}$$
$$= \gamma \sum_s \boldsymbol{x}(s) \sum_{a,s'} \boldsymbol{P}_\mu(s' \mid s, a) \nabla_\theta \pi(a \mid s) \, V^\pi_\mu(s')$$
$$= \gamma \sum_s \boldsymbol{x}(s) \sum_a \nabla_\theta \pi(a \mid s) \sum_{s'} \boldsymbol{P}_\mu(s' \mid s, a) V^\pi_\mu(s')$$
$$= \gamma \sum_s \boldsymbol{x}(s) \sum_a \nabla_\theta \pi(a \mid s) \, Q^\pi_\mu(s, a).$$

The last claim (8.4.5) is straightforward. Indeed,

$$\nabla_\theta \mathbb{E}\, U = \sum_h U(h) \nabla_\theta \mathbb{P}^\pi_\mu(h) = \sum_h U(h) \, \mathbb{P}^\pi_\mu(h) \nabla \ln \mathbb{P}^\pi_\mu(h), \tag{8.4.6}$$

as $\nabla_\theta \ln \mathbb{P}^\pi_\mu(h) = \frac{1}{\mathbb{P}^\pi_\mu(h)} \nabla_\theta \mathbb{P}^\pi_\mu(h)$. □

8.4.1 Stochastic Policy Gradient

For finite MDPs, we can obtain $\boldsymbol{x}_\pi$ from the state occupancy matrix (6.5.2) by left multiplication with the initial state distribution $\boldsymbol{y}$. However, in the context of gradient methods, it makes more sense to use a stochastic estimate of $\boldsymbol{x}_\pi$ to calculate the gradient, since

$$\nabla_\theta \mathbb{E}^\pi U = \mathbb{E}^\pi_y \sum_{s,t} \gamma^t \, \mathbb{I}\{s_t = s\} \sum_a \nabla_\theta \pi(a \mid s) \, Q^\pi(s, a).$$

For the discounted reward criterion, we can easily obtain unbiased samples through geometric stopping (see Exercise 6.1).

Importance sampling
The last formulation is especially useful as it allows us to use importance sampling to compute the gradient even on data obtained for different policies, which in general is more data efficient. First note that for any history $h \in (\mathcal{S} \times \mathcal{A})^*$, we have

$$\mathbb{P}_\mu^\pi(h) = \prod_{t=1}^T \mathbb{P}_\mu(s_t \mid s^{t-1}, a^{t-1})\, \mathbb{P}^\pi(a_t \mid s^t, a^{t-1}) \tag{8.4.7}$$

without any Markovian assumptions on the model or policy. We can now rewrite (8.4.6) in terms of the expectation with respect to an alternative policy π' as

$$\begin{aligned}\nabla\, \mathbb{E}_\mu^\pi U &= \mathbb{E}_\mu^{\pi'} \left(U(h) \nabla \ln \mathbb{P}_\mu^\pi(h) \frac{\mathbb{P}_\mu^\pi(U)}{\mathbb{P}_\mu^{\pi'}(U)} \right) \\ &= \mathbb{E}_\mu^{\pi'} \left(U(h) \nabla \ln \mathbb{P}_\mu^\pi(h) \prod_{t=1}^T \frac{\pi(a_t \mid s^t, a^{t-1})}{\pi'(a_t \mid s^t, a^{t-1})} \right),\end{aligned}$$

since the μ-dependent terms in (8.4.7) cancel out. In practice the expectation would be approximated through sampling trajectories h. Note that

$$\nabla \ln \mathbb{P}_\mu^\pi(h) = \sum_t \nabla \ln \pi(a_t \mid s^t, a^{t-1}) = \sum_t \frac{\nabla \pi(a_t \mid s^t, a^{t-1})}{\pi(a_t \mid s^t, a^{t-1})}.$$

Overall, importance sampling allows us to perform stochastic gradient descent on data collected from any arbitrary previous policy π', and perform gradient descent on a parametrized policy π.

8.4.2 Practical Considerations

The first design choice in any gradient algorithm is how to parametrize the policy. For the discrete case, a common parametrization is to have a separate and independent parameter for each state-action pair, i.e., $\theta_{s,a} = \pi(a|s)$. This leads to a particularly simple expression for the second form (8.4.4), which is $\partial/\partial_{\theta_{s,a}} \mathbb{E}_\mu^\pi U = y(s)\, Q(s, a)$. However, it is easy to see that in this case the parametrization will lead to all parameters increasing if rewards are positive. This can be avoided by either a Softmax parametrization or by subtracting a bias term (e.g. the value of the state $V_\mu^\pi(s)$) from the derivative. Nevertheless, this parametrization implies stochastic discrete policies.

We could also suitably parametrize continuous policies. For example, if $\mathcal{A} \subset \mathbb{R}^n$, we can consider a linear policy. Most of the derivation carries over to Euclidean state-action spaces. In particular, the second form (8.4.4) is also suitable for deterministic policies.

Finally, in practice, we may not need to accurately calculate the expectations involved in the gradient. Sample trajectories are sufficient to update the gradient in a meaningful way, especially for the third form (8.4.5), as we can naturally sample from the distribution of trajectories. However, the fact that this form doesn't need a Markovian assumption also means that it cannot take advantage of Markovian environments.

Policy gradient methods are useful, especially in cases where the environment model or value function is extremely complicated, while the optimal policy itself might be quite simple. The main difficulty lies in obtaining an appropriate estimate of the gradient itself, but convergence to a local maximum is generally good as long as we are adjusting the parameters in a gradient-related direction (in the sense of Assumption 7.1.1iii).

8.5 Examples

Let us now consider two well-known problems with a 2-dimensional continuous state space and a discrete set of actions. The first is the *inverted pendulum* problem in the version of Lagoudakis and Parr [8], where a controller must balance a rod upside-down. The state information is the rotational velocity and position of the pendulum. The second example is the *mountain car* problem, where we must drive an underpowered vehicle to the top of a hill [1]. The state information is the velocity and location of the car. In both problems, there are three actions: "push left", "push right" and "do nothing".

Let us first consider the effect of model and features in representing the value function of the inverted pendulum problem. Figure 8.5 shows value function approximations for policy evaluation under a uniformly random policy for different choices of model and features. Here we need to fit an approximate value function to samples of the utility obtained from different states. The quality of the approximation depends on both the model and the features used. The first choice of features is simply the raw state representation, while the second a 16×16 uniform RBF tiling. The two models are very simple: the first uses a linear model-Gaussian model[4] assumption on observation noise (LG), and the second is a k-nearest neighbour (kNN) model.

As can be seen from the figure the linear model results in a smooth approximation, but is inadequate for modelling the value function in the original 2-dimensional state space. However, a high-dimensional non-linear projection using RBF kernels results in a smooth and accurate value function representation. Non-parametric models such as k-nearest neighbours behave rather well under either state representation.

---

[4] Essentially, this is the a linear model of the form $s_{t+1} \mid s_t = s, a_t = a \sim \mathcal{N} \mu_a^\top s, \Sigma_a$, where μ_a has a normal prior and Σ a Wishart prior.

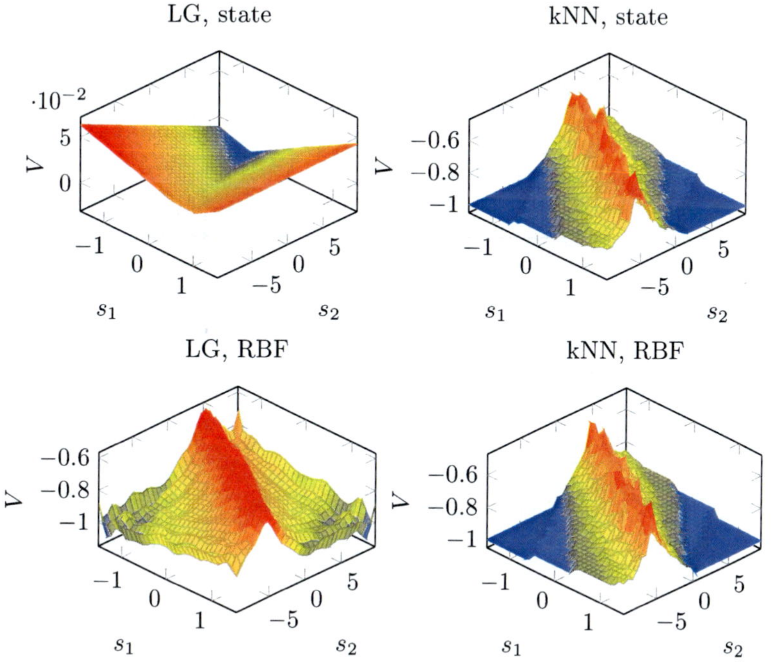

Fig. 8.5 Estimated value function of a uniformly random policy on the 2-dimensional state-space of the pendulum problem. Results are shown for a k-nearest neighbour model (kNN) with $k = 3$ and a Bayesian linear-Gaussian model (LG) for either the case when the model uses the plain state information (state) or an 256-dimensional RBF embedding (RBF)

For finding the optimal value function we must additionally consider the question of which algorithm to use. In Fig. 8.6 we see the effect of choosing either approximate value iteration (AVI) or representative state representations and value iteration (RSVI) for the inverted pendulum and mountain car.

8.6 Further Reading

Among value function approximation methods, the two most well known are fitted Q-iteration [10] and fitted value iteration, which has been analysed in [11]. Minimizing the Bellman error [12–14] is generally a good way to ensure that approximate value iteration is stable.

In approximate policy iteration methods, one needs to approximate both the value function and policy. An empirical approximation of the value function is maintained in rollout sampling policy iteration [5, 6]. However, one can employ least-squares methods [7, 8, 15] for example.

The general technique of state aggregation [16, 17] is applicable to a variety of reinforcement learning algorithms. While the more general question of selecting

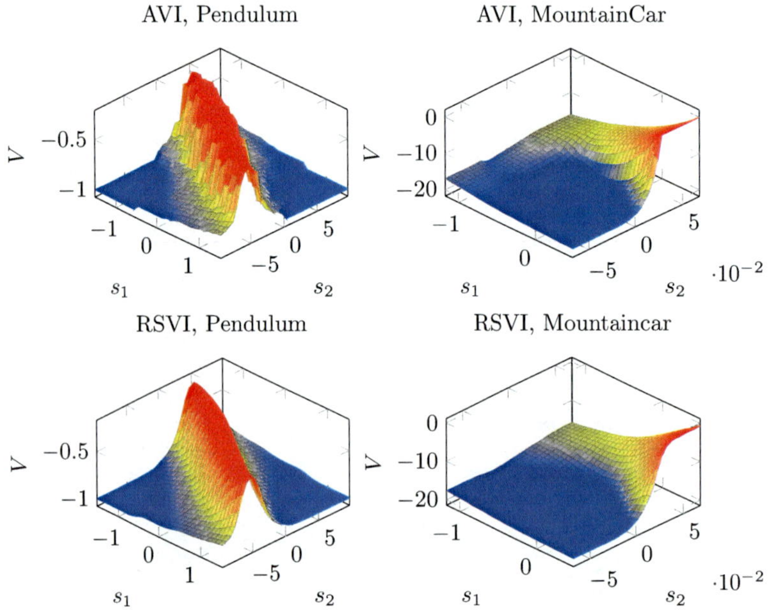

Fig. 8.6 Estimated optimal value function for the pendulum problem. Results are shown for approximate value iteration (AVI) with a Bayesian linear-Gaussian model, and a representative state representation (RSVI) with an RBF embedding. Both the embedding and the states where the value function is approximated are a 16×16 uniform grid over the state space

appropriate features is open, there has been some progress in the domain of feature reinforcement learning [18]. In general, learning internal representations (i.e., features) has been a prominent aspect of neural network research [19]. Even if it is unclear to what extent recently proposed approximation architectures that employ deep learning actually learn useful representations, they have been successfully used in combination with simple reinforcement learning algorithms [20]. Another interesting direction is to establish links between features and approximately sufficient statistics [21, 22].

Finally, the policy gradient theorem in the state visitation form was first proposed by Sutton et al. [9], while Williams [23] was the first to use the log-ratio trick in Eq. (8.4.5) in reinforcement learning. To our knowledge, the analytical gradient has not actually been applied (or indeed, described) in prior literature. Extensions of the policy gradient idea are also natural. They have also been used in a Bayesian setting by Ghavamzadeh and Engel [14], while the natural gradient has been proposed by Kakade [24]. A survey of policy gradient methods can be found in [25].

8.7 Exercises

Exercise 8.1 (Enlarging the function space) Consider the problem in Example 8.1. What would be a simple way to extend the space of value functions from the three given candidates to an infinite number of value functions? How could we get a good fit?

Exercise 8.2 (Enlarging the policy) Consider Example 8.2. This represents an example of linear deterministic policies. In which two ways can this policy space be extended and how?

Exercise 8.3 Find the derivative for minimizing the cost function in (8.1.2) for the following two cases:

1. $p = 2, \kappa = 2$.
2. $p \to \infty, \kappa = 1$.

References

1. Sutton, R.S., Barto, A.G.: Reinforcement Learning: An Introduction. MIT Press (1998)
2. Lagoudakis, M.G., Parr, R.: Reinforcement learning as classification: leveraging modern classifiers. In: Machine Learning, Proceedings of the 20th International Conference (ICML 2003), pp. 424–431. AAAI Press (2003)
3. Even-Dar, E., Mansour, Y: Approximate equivalence of Markov decision processes. In: Computational Learning Theory and Kernel Machines, 16th Annual Conference on Computational Learning Theory and 7th Kernel Workshop, COLT/Kernel 2003. Lecture notes in Computer Science, vol. 2777, pp. 581–594. Springer (2003)
4. Bertsekas, D.P., Tsitsiklis, J.N.: Neuro-Dynamic Programming. Athena Scientific (1996)
5. Dimitrakakis, C., Lagoudakis, M.G.: Rollout sampling approximate policy iteration. Mach. Learn. **72**(3), 157–171 (2008)
6. Dimitrakakis, C., Lagoudakis, M.G.: Algorithms and bounds for rollout sampling approximate policy iteration. In: Girgin, S., Loth, M., Munos, R., Preux, P., Ryabko, D. (eds.) Recent Advances in Reinforcement Learning, 8th European Workshop, EWRL 2008. Lecture Notes in Computer Science, vol. 5323, pp. 27–40. Springer (2008)
7. Bradtke, Steven J., Barto, Andrew G.: Linear least-squares algorithms for temporal difference learning. Mach. Learn. **22**(1), 33–57 (1996)
8. Lagoudakis, Michail G.: Parr, Ronald: least-squares policy iteration. J. Mach. Learn. Res. **4**, 1107–1149 (2003)
9. Sutton, R.S., McAllester, D.A., Singh, S.P., Mansour, Y.: Policy gradient methods for reinforcement learning with function approximation. In: Advances in Neural Information Processing Systems 12, pp. 1057–1063. The MIT Press (1999)
10. Antos, A., Munos, R., Szepesvari, C.: Fitted Q-iteration in continuous action-space MDPs. In: Advances in Neural Information Processing Systems, vol. 20, pp. 9–16 (2008)
11. Munos, R., Szepesvári, C.: Finite-time bounds for fitted value iteration. J. Mach. Learn. Res. **9**, 815–857 (2008)
12. Antos, A., Szepesvari, C., Munos, R.: Learning near-optimal policies with bellman-residual minimization based fitted policy iteration and a single sample path. Mach. Learn. **71**(1), 89–129 (2008)

13. Dimitrakakis, C.: Monte-carlo utility estimates for bayesian reinforcement learning. In: Proceedings of the 52nd IEEE Conference on Decision and Control, CDC 2013, pp. 7303–7308. IEEE (2013)
14. Ghavamzadeh, M., Engel, Y.: Bayesian policy gradient algorithms. In: Advances in Neural Information Processing Systems, vol. 19, pp. 457–464. MIT Press (2006)
15. Boyan, Justin A.: Technical update: least-squares temporal difference learning. Mach. Learn. **49**(2), 233–246 (2002)
16. Singh, Satinder P., Jaakkola, Tommi S., Jordan, Michael I.: Reinforcement learning with soft state aggregation. Adv. Neural Inf. Process. Syst. **7**, 361–368 (1995)
17. Bernstein, A.: Adaptive state aggregation for reinforcement learning. Master's thesis, Technion Israel Institute of Technology (2007)
18. Hutter, Marcus: Feature reinforcement learning: part I: unstructured MDPs. J. Artif. General Intell. **1**, 3–24 (2009)
19. Rumelhart, D.E., Hinton, G.E., Williams, R.J.: Learning internal representations by error propagation. In: Rumelhart, D.E., McClelland, J.L., et al. (eds.) Parallel Distributed Processing. Foundations, vol. 1, pp. 318–362. MIT Press, Cambridge (1987)
20. Mnih, V., Kavukcuoglu, K., Silver, D., et al.: Human-level control through deep reinforcement learning. Nature **518**(7540), 529–533 (2015)
21. Dimitrakakis, C., Tziortziotis, N.: ABC reinforcement learning. In: Proceedings of the 30th International Conference on Machine Learning, ICML 2013, pp. 684–692. JMLR.org (2013)
22. Dimitrakakis, C., Tziortziotis, N.: Usable ABC reinforcement learning. In: NIPS 2014 Workshop: ABC in Montreal (2014)
23. Williams, Ronald J.: Simple statistical gradient-following algorithms for connectionist reinforcement learning. Mach. Learn. **8**(3–4), 229–256 (1992)
24. Kakade, Sham: A natural policy gradient. Adv. Neural Inf. Process. Syst. **14**, 1531–1538 (2002)
25. Peters, J., Schaal, S.: Policy gradient methods for robotics. In: 2006 IEEE/RSJ International Conference on Intelligent Robots and Systems, IROS 2006, pp. 2219–2225. IEEE (2006)

Chapter 9
Bayesian Reinforcement Learning

9.1 Introduction

In this chapter, we return to the setting of subjective probability and utility by formalizing the reinforcement learning problem as a Bayesian decision problem and solving it directly. In the Bayesian setting, we are acting in an unknown environment, and we represent our subjective belief about the environment in the form of a probability distribution. We shall first consider the case of acting in unknown MDPs, which is the focus of the reinforcement learning problem. We will examine a few different heuristics for maximizing expected utility in the Bayesian setting and contrast them with tractable approximations to the Bayes-optimal solution. We then present extensions of these ideas to continuous domains. Finally, we draw connections of this setting to partially observable MDPs.

9.1.1 Acting in Unknown MDPs

The reinforcement learning problem can be formulated as the problem of learning to act in an unknown environment, only by interaction and reinforcement. All of these elements of the definition are important. Firstly and foremostly it is a *learning* problem: We have only partial prior knowledge about the environment we are acting in. This knowledge is arrived at via *interaction* with the environment. We do not have a fixed set of data to work with, but we must actively explore the environment to understand how it works. Finally, there is a *reinforcement* signal that punishes some behaviours and rewards others. We can formulate some of these problems as Markov decision processes.

Let us begin with the case where the environment can be represented as an MDP μ. That is, at each time step t, we observe the environment's state $s_t \in \mathcal{S}$, take an action $a_t \in \mathcal{A}$ and receive *reward* $r_t \in \mathbb{R}$. Consequently, the state and our action fully

C. Dimitrakakis and R. Ortner, *Decision Making Under Uncertainty and Reinforcement Learning*, Intelligent Systems Reference Library 223,
https://doi.org/10.1007/978-3-031-07614-5_9

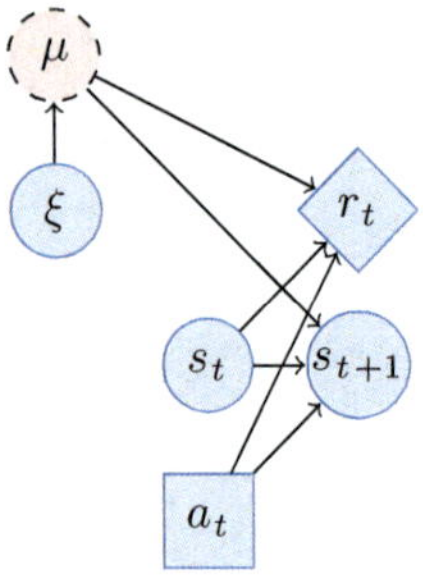

Fig. 9.1 The unknown Markov decision process. ξ is our prior over the unknown μ, which is not directly observed. However, we always observe the result of our action a_t in terms of reward r_t and next state s_{t+1}

determine the distribution of the immediate reward, as well as that of the next state, as described in Definition 6.3.1. For a specific MDP μ the probability of the immediate reward is given by $\mathbb{P}_\mu(r_t \mid s_t, a_t)$, with expectation $\bar{r}_\mu(s, a) \triangleq \mathbb{E}_\mu(r_t \mid s_t = s, a_t = a)$, while the next state distribution is given by $\mathbb{P}_\mu(s_{t+1} \mid s_t, a_t)$. If these quantities are known, or if we can at least draw samples from these distributions, it is possible to employ (approximate) dynamic programming to obtain the optimal policy and value function for the MDP.

More precisely, when μ is known, we wish to find a *policy* $\pi : \mathcal{S} \to \mathcal{A}$ maximizing the *utility* in expectation. This requires us to solve the maximization problem $\max_\pi \mathbb{E}^\pi_\mu U$, where the utility is an additive function of rewards, $U = \sum_{t=1}^T r_t$. This can be accomplished using standard algorithms, such as value or policy iteration. However, knowing μ is contrary to the problem definition.

In Chap. 7 we have seen a number of stochastic approximation algorithms which allow us to learn the optimal policy for a given MDP eventually. However, these generally give few guarantees on the performance of the policy while learning. A good way of learning the optimal policy in an MDP should trade off exploring the environment to obtain further knowledge and simultaneously exploiting this knowledge.

Within the subjective probabilistic framework, there is a natural formalization for learning optimal behavior in am MDP. We define a prior belief ξ on the set of MDPs $\mathcal{M}$, and then find the policy that maximizes the expected utility with respect to the prior $\mathbb{E}^\pi_\xi(U)$. The structure of the unknown MDP process is shown in Fig. 9.1. We have previously seen two simpler sequential decision problems in the Bayesian setting. The first was the simple optimal stopping procedure in Sect. 5.2.2, which introduced the backwards induction algorithm. The second was the optimal experiment design problem, which resulted in the bandit Markov decision process of Sect. 6.2. Now we want to formulate the reinforcement learning problem as a Bayesian maximization problem.

Let ξ be a prior over $\mathcal{M}$ and Π be a set of policies. Then the expected utility of the optimal policy, over some fixed starting state distribution, is

$$U^*_\xi \triangleq \max_{\pi \in \Pi} \mathbb{E}(U \mid \pi, \xi) = \max_{\pi \in \Pi} \int_{\mathcal{M}} \mathbb{E}(U \mid \pi, \mu) \, \mathrm{d}\xi(\mu). \tag{9.1.1}$$

Solving this optimization problem and hence finding the optimal policy is however not easy, as in general the optimal policy π must incorporate the information it obtained while interacting with the MDP. Formally, this means that it must map from *histories* to actions. For any such history-dependent policy, the action we take at step t must depend on what we observed in previous steps $1, \ldots, t-1$. Consequently, an optimal policy must also specify actions to be taken in all future time steps and accordingly take into account the learning that will take place up to each future time step. Thus, in some sense, the value of information is automatically taken into account in this model. This is illustrated in the following example.

Example 9.1 Consider two MDPs μ_1, μ_2 with a single state (i.e., $\mathcal{S} = \{1\}$) and actions $\mathcal{A} = \{1, 2\}$. In the MDP μ_i, whenever you take action $a_t = i$ you obtain reward $r_t = 1$, otherwise you obtain reward 0. If we only consider policies that do not take into account the history so far, the expected utility of such a policy π taking action i with probability $\pi(i)$ is

$$\mathbb{E}_\xi^\pi U = T \sum_i \xi(\mu_i)\, \pi(i)$$

for horizon T. Consequently, if the prior ξ is not uniform, the optimal policy selects the action corresponding to the MDP with the highest prior probability. Then, the maximal expected utility is

$$T \max_i \xi(\mu_i).$$

However, observing the reward after choosing the first action, we can determine the true MDP. Consequently, an improved policy is the following: First select the best action with respect to the prior, and then switch to the best action for the MDP we have identified to be the true one. Then, our utility improves to

$$\max_i \xi(\mu_i) + (T - 1).$$

As we have to consider quite general policies in this setting, it is useful to differentiate between different policy types. We use Π to denote the set of all policies. We use Π_k to denote the set of k-order Markov policies π, that only take into account the previous k steps, that is,

$$\pi(a_t \mid s^t, a^{t-1}, r^{t-1}) = \pi(a_t \mid s_{t-k+1}^t, a_{t-k}^{t-1}, r_{t-k}^{t-1}),$$

where here and in the following we use the notation s^t to abbreviate $(s_1, \ldots, s_t)$ and s_t^{t+k} for $(s_t, \ldots, s_{t+k})$, and accordingly a_t, r_t, a_t^{t+k}, and r_t^{t+k}. Important special cases are the set of *blind* policies Π_0 and the set of *memoryless* policies Π_1. The set $\bar{\Pi}_k \subset \Pi_k$ contains all *stationary* policies in Π_k. Finally, policies may be indexed by some parameter set Θ, in which case the set of parameterized policies is given by Π_Θ.

Let us now turn to the problem of learning an optimal policy. Learning means that observations we make will affect our belief, so that we will first take a closer look at this belief update. Given that, we shall examine methods for exact and approximate methods of policy optimization.

9.1.2 Updating the Belief

Strictly speaking, in order to update our belief, we must condition the prior distribution on all the information. This includes the sequence of observations up to this point in time, including the states s^t, actions a^{t-1}, and rewards r^{t-1}, as well the policy π that we followed. Let $D_t = \langle s^t, a^{t-1}, r^{t-1} \rangle$ be the observed data up to time t. Then the posterior measure for any measurable subset B of the set of all MDPs $\mathcal{M}$ is

$$\xi(B \mid D_t, \pi) = \frac{\int_B \mathbb{P}^\pi_\mu(D_t)\, \mathrm{d}\xi(\mu)}{\int_{\mathcal{M}} \mathbb{P}^\pi_\mu(D_t)\, \mathrm{d}\xi(\mu)}.$$

However, as we shall see in the following remark, we can usually[1] ignore the policy itself when calculating the posterior.

Remark 9.1.1 The dependence on the policy can be removed, since the posterior is the same for all policies that put non-zero mass on the observed data. Indeed, for $D_t \sim \mathbb{P}^\pi_\mu$ it is easy to see that $\forall \pi' \neq \pi$ such that $\mathbb{P}^{\pi'}_\mu(D_t) > 0$, it holds that

$$\xi(B \mid D_t, \pi) = \xi(B \mid D_t, \pi').$$

The proof is left as an exercise for the reader. In the specific case of MDPs, the posterior calculation is easy to perform incrementally. This also more clearly demonstrates why there is no dependence on the policy. Let ξ_t be the (random) posterior at time t. Then the next-step belief is given by

$$\begin{aligned}
\xi_{t+1}(B) \triangleq \xi(B \mid D_{t+1}) &= \frac{\int_B \mathbb{P}^\pi_\mu(D_t)\, \mathrm{d}\xi(\mu)}{\int_{\mathcal{M}} \mathbb{P}^\pi_\mu(D_t)\, \mathrm{d}\xi(\mu)} \\
&= \frac{\int_B \mathbb{P}_\mu(s_{t+1}, r_t \mid s_t, a_t)\, \pi(a_t \mid s^t, a^{t-1}, r^{t-1})\, \mathrm{d}\xi(\mu \mid D_t)}{\int_{\mathcal{M}} \mathbb{P}_\mu(s_{t+1}, r_t \mid s_t, a_t)\, \pi(a_t \mid s^t, a^{t-1}, r^{t-1})\, \mathrm{d}\xi(\mu \mid D_t)} \\
&= \frac{\int_B \mathbb{P}_\mu(s_{t+1}, r_t \mid s_t, a_t)\, \mathrm{d}\xi_t(\mu)}{\int_{\mathcal{M}} \mathbb{P}_\mu(s_{t+1}, r_t \mid s_t, a_t)\, \mathrm{d}\xi_t(\mu)}.
\end{aligned}$$

[1] The exception involves any type of inference where $\mathbb{P}^\pi_\mu(D_t)$ is not directly available. This includes methods of approximate Bayesian computation [1], that use trajectories from past policies for approximation. See [2] for an example of this in reinforcement learning.

The above calculation is easy to perform for arbitrarily complex MDPs when the set $\mathcal{M}$ is finite. The posterior calculation is also simple under certain conjugate priors, such as the Dirichlet-multinomial prior for transition distributions.

9.2 Finding Bayes-Optimal Policies

The problem of policy optimization in the Bayesian case is much harder than when the MDP is known. This is simply because we have to consider history dependent policies, which makes the policy space much larger.

In this section, we first consider two simple heuristics for finding sub-optimal policies. Then we show how policy gradient methods can be extended to the Bayesian case to obtain Bayes-optimal policies within a parametrized policy class. We proceed considering finite look-ahead backwards induction to approximate the Bayes-optimal policy. Generally, backwards induction in this setting requires building an exponential-size tree. However, upper and lower bounds on the value function can be used to create a *branch and bound* algorithm to improve efficiency. We end this section introducing two methods to construct such bounds, and discuss their relation to one of the best-known Bayesian methods, *posterior sampling*.

9.2.1 The Expected MDP Heuristic

One simple heuristic is to simply calculate the expected MDP $\widehat{\mu}(\xi) \triangleq \mathbb{E}_\xi \mu$ for the current belief ξ. In particular, the transition kernel of the expected MDP is simply the expected transition kernel:

$$\mathbb{P}_{\widehat{\mu}(\xi)}(s'|s, a) = \int_{\mathcal{M}} \mathbb{P}_\mu(s'|s, a) \, \mathrm{d}\xi(\mu)$$

Then we simply calculate the optimal memoryless policy for $\widehat{\mu}(\xi)$, that is,

$$\pi^*(\widehat{\mu}(\xi)) \in \arg\max_{\pi \in \Pi_1} V^\pi_{\widehat{\mu}(\xi)},$$

where $\Pi_1 = \left\{\pi \in \Pi \mid \mathbb{P}^\pi(a_t \mid s^t, a^{t-1}) = \mathbb{P}^\pi(a_t \mid s_t)\right\}$ is the set of Markov policies. The policy $\pi^*(\widehat{\mu}(\xi))$ is executed on the real MDP. Algorithm 9.1 shows the pseudocode for this heuristic. One important detail is that the policy update schedule only generates the k-th policy at step T_k. This is useful to ensure policies remain consistent, as small changes in the mean MDP may create a large change in the resulting policy. It is natural to have $T_k - T_{k-1}$ in the order of $1/1 - \gamma$ for discounted problems, or simply the length of the episode for episodic problems. In the undiscounted case,

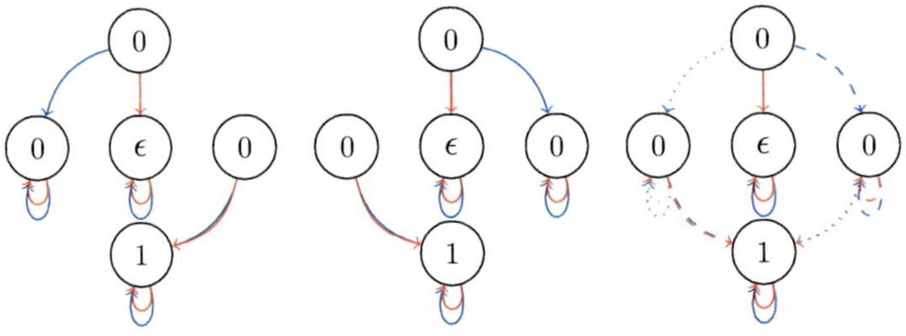

(a) μ_1 (b) μ_2 (c) $\widehat{\mu}(\xi)$

Fig. 9.2 The two MDPs and the expected MDP from Example 9.2

switching policies whenever sufficient information has been obtained to significantly change the belief gives good performance guarantees, as we shall see in Chap. 10.

Algorithm 9.1 The expected MDP heuristic

input prior ξ_0, update schedule $\{T_k\}$
for $k = 1, \ldots$ **do**
  $\pi_k \approx \pi^*(\widehat{\mu}(\xi)) \in \arg\max_{\pi \in \Pi_1} V^{\pi}_{\widehat{\mu}(\xi)}$
  for $t = T_{k-1} + 1, \ldots, T_k$ **do**
    Observe s_t.
    Update belief $\xi_t(\cdot) = \xi_{t-1}(\cdot \mid s_t, a_{t-1}, r_{t-1}, s_{t-1})$.
    Take action $a_t \sim \pi_k(\cdot \mid s_t)$.
    Observe reward r_t.
  end for
end for

Unfortunately, the policy returned by this heuristic may be far from the Bayes-optimal policy in Π_1, as shown by the following example.

Example 9.2 (*Counterexample to Algorithm 9.1 based on an example by Remi Munos*) As illustrated in Fig. 9.2, let $\mathcal{M} = \{\mu_1, \mu_2\}$ be the set of MDPs, and the belief is $\xi(\mu_1) = \theta$, $\xi(\mu_2) = 1 - \theta$. All transitions are deterministic, and there are two actions, the blue and the red action. We see that in the expected MDP the state with reward 1 is reachable, and that $\widehat{\mu}(\xi) \notin \mathcal{M}$. One can compute that even when $T \to \infty$, the $\widehat{\mu}(\xi)$-optimal policy is not optimal in Π_1, if

$$\epsilon < \frac{\gamma\theta(1-\theta)}{1-\gamma}\left(\frac{1}{1-\gamma\theta} + \frac{1}{1-\gamma(1-\theta)}\right).$$

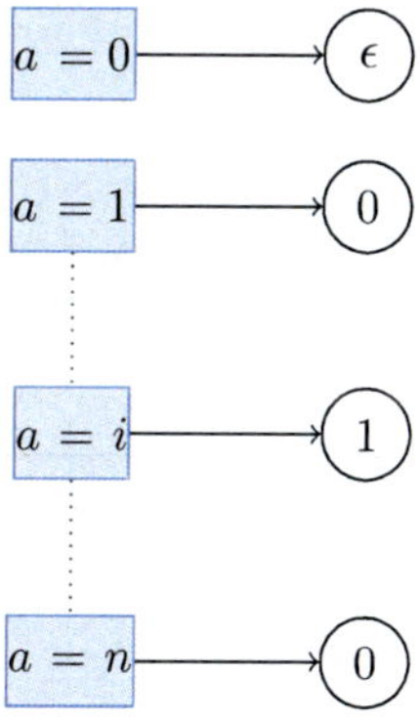

Fig. 9.3 The MDP μ_i from Example 9.3

9.2.2 *The Maximum MDP Heuristic*

An alternative idea is to simply pick the maximum-probability MDP, as shown in Algorithm 9.2. This at least guarantees that the MDP for which one chooses the optimal policy is actually within the set of MDPs. However, it may still be the case that the resulting policy is sub-optimal, as shown by the following example.

Algorithm 9.2 The maximum MDP heuristic

input prior ξ_0, update schedule $\{T_k\}$
for $k = 1, \dots$ **do**
  $\pi_k \approx \arg\max_\pi \mathbb{E}^\pi_{\hat{\mu}^*(\xi_{T_k})} U$.
  for $t = 1 + T_{k-1}, \dots, T_k$ **do**
    Observe s_t.
    Update belief $\xi_t(\cdot) = \xi_{t-1}(\cdot \mid s_t, a_{t-1}, r_{t-1}, s_{t-1})$.
    Take action $a_t \sim \pi_k(\cdot \mid s_t)$.
    Observe reward r_t.
  end for
end for

Example 9.3 (*Counterexample to Algorithm 9.2*) As illustrated in Fig. 9.3 let $\mathcal{M} = \{\mu_i \mid i = 1, \dots, n\}$ be the set of MDPs where $\mathcal{A} = \{0, \dots, n\}$. In all MDPs, action 0 gives a reward of ϵ. In MDP each μ_i, action i gives reward 1, and all remaining actions give a reward of 0. For any action a, the MDP terminates after an action is chosen and the reward received. Now if $\xi(\mu_i) < \epsilon$ for all i, then it is optimal to choose action 0, while Algorithm 9.2 would pick the sub-optimal $\max_i \xi(\mu_i)$.

9.2.3 Bayesian Policy Gradient

Policy gradient (see Sect. 8.4) can also be performed in the Bayesian setting. For this, we must restrict our policies to a parametrized policy space so that we can differentiate them:

$$\Pi_\Theta \triangleq \{\pi_\theta \mid \theta \in \Theta\}$$

The general idea is to find the policy parameter maximizing expected utility under the current belief, i.e. solve the problem

$$\max_\theta \mathbb{E}^\pi_\xi[U],$$

where the utility is defined with respect to some starting state distribution (see Eq. 8.4.2). A policy gradient algorithm would simply move in the direction of the gradient, which can be computed as

$$\begin{aligned}
\nabla_\theta \mathbb{E}^\pi_\xi[U] &= \nabla_\theta \int_{\mathcal{M}} \mathbb{E}^\pi_\mu[U]\, \mathrm{d}\xi(\mu) \\
&= \int_{\mathcal{M}} \nabla_\theta \mathbb{E}^\pi_\mu[U]\, \mathrm{d}\xi(\mu) \\
&\approx \frac{1}{n}\sum_{i=1}^{n} \nabla_\theta \mathbb{E}^\pi_{\mu^{(i)}}[U], \qquad\qquad \mu^{(i)} \sim \xi(\mu).
\end{aligned}$$

Here, the integral is approximated by sampling MDPs from the belief, and $\nabla_\theta \mathbb{E}^\pi_{\mu^{(i)}}[U]$ is the standard policy gradient for a given MDP $\mu^{(i)}$. Approximations to the gradient can be computed via rollouts.

An interesting question is how to define the policy parametrization. In order for a policy to be adaptive it must take into account the complete history of observations. This can be achieved even for simple policies, as long as we have a statistic mapping histories to a rich enough representation. This is detailed in the following example.

Example 9.4 (*Policies defined on a statistic* $\phi : \mathcal{H} \times \mathcal{A} \to \mathbb{R}^{k\times|A|}$) Define history $h_t = s_1, a_1, r_1, \ldots, s_t$ and a history-dependent sigmoid policy:

$$\pi_\theta(a \mid h_t) \propto \exp\left(\phi(h_t, a)^\top \theta\right)$$

9.2.4 The Belief-Augmented MDP

The most direct way to actually solve the Bayesian reinforcement learning problem of (9.1.1) is to cast it as a yet another MDP. We have already seen how this can be done with bandit problems in Sect. 6.2.2, but we shall now see that the general methodology is also applicable to MDPs.

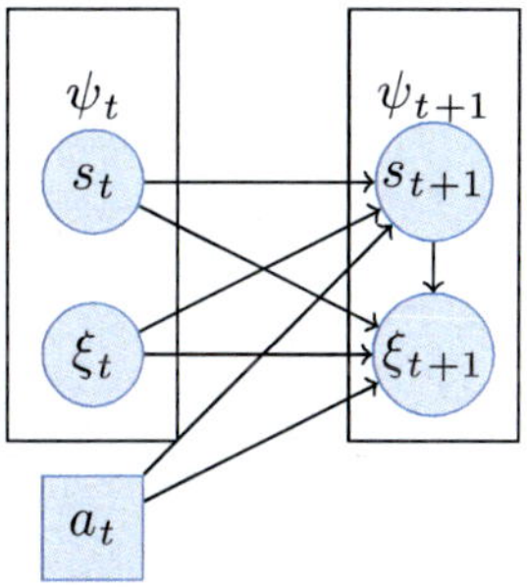

(a) The complete MDP model.

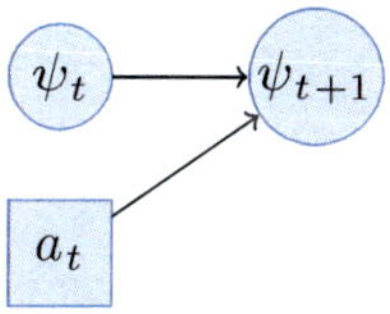

(b) Compact form of the model.

Fig. 9.4 Belief-augmented MDP

We are given an initial belief ξ_0 on a set of MDPs $\mathcal{M}$. Each $\mu \in \mathcal{M}$ is a tuple $(\mathcal{S}, \mathcal{A}, P_\mu, \rho)$, with state space $\mathcal{S}$, action space $\mathcal{A}$, transition kernel P_μ and reward function $\rho : \mathcal{S} \times \mathcal{A} \to \mathbb{R}$. Let s_t, a_t, r_t be the state, action, and reward observed in the original MDP and ξ_t be our belief over MDPs $\mu \in \mathcal{M}$ at step t. Note that the marginal next-state distribution is

$$P(s_{t+1} \in S \mid \xi_t, s_t, a_t) \triangleq \int_{\mathcal{M}} P_\mu(s_{t+1} \in S \mid s_t, a_t) \, \mathrm{d}\xi_t(\mu),$$

while the next belief deterministically depends on the next state, i.e.,

$$\xi_{t+1}(\cdot) \triangleq \xi_t(\cdot \mid s_{t+1}, s_t, a_t).$$

We now construct an augmented Markov decision process $(\Psi, \mathcal{A}, P, \rho')$ with the state space $\Psi = \mathcal{S} \times \Xi$ being the product of the original MDP states $\mathcal{S}$ and possible beliefs Ξ. The transition distribution is given by

$$\begin{aligned} P(\psi_{t+1} \mid \psi_t, a_t) &= P(\xi_{t+1}, s_{t+1} \mid \xi_t, s_t, a_t) \\ &= P(\xi_{t+1} \mid \xi_t, s_{t+1}, s_t, a_t) P(s_{t+1} \mid \xi_t, s_t, a_t), \end{aligned}$$

where $P(\xi_{t+1} \mid \xi_t, s_{t+1}, s_t, a_t)$ is the singular distribution centred on the posterior distribution $\xi_t(\cdot \mid s_{t+1}, s_t, a_t)$.

This construction is illustrated in Fig. 9.4. The optimal policy for the augmented MDP is the ξ-optimal policy in the original MDP. The augmented MDP has a pseudo-tree structure (since belief states might repeat), as shown in the following example.

Example 9.5 Consider aset of MDPs $\mathcal{M}$ with $\mathcal{A} = \{1, 2\}$, $\mathcal{S} = \{1, 2\}$. In general for any hyper-state $\psi_t = (s_t, \xi_t)$ each possible action-state transition results in one specific new hyper-state. This is illustrated for the specific example in the following diagram.

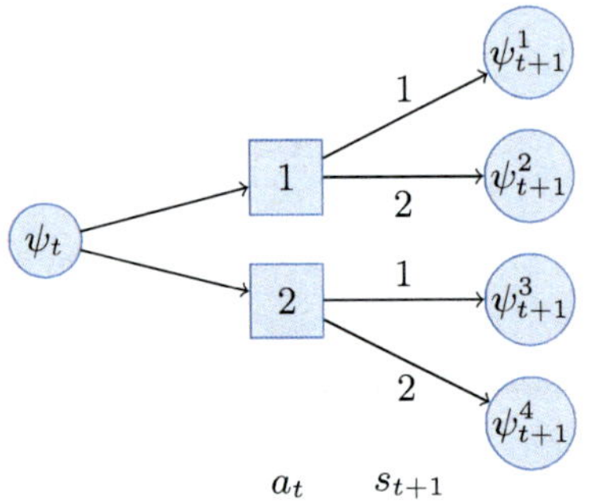

When the branching factor is very large, or when we need to deal with very large tree depths, it becomes necessary to approximate the MDP structure.

9.2.5 Branch and Bound

Branch and bound is a general technique for solving large problems. It can be applied in all cases where upper and lower bounds on the value of solution sets can be found. For Bayesian reinforcement learning, we can consider upper and lower bounds $\boldsymbol{q}^+$ and $\boldsymbol{q}^-$ on Q^* in the belief-augmented MDP (BAMDP). That is,

$$\boldsymbol{q}^+(\psi, a) \geq Q^*(\psi, a) \geq \boldsymbol{q}^-(\psi, a)$$
$$\boldsymbol{v}^+(\psi) = \max_{a \in \mathcal{A}} \boldsymbol{q}^+(\psi, a), \qquad \boldsymbol{v}^-(\psi) = \max_{a \in \mathcal{A}} \boldsymbol{q}^-(\psi, a).$$

Let us now consider an incremental expansion of the BAMDP so that, starting from some hyperstate ψ_t, we create the BAMDP tree for all subsequent states $\psi_{t+1}, \psi_{t+2}, \ldots$. For any leaf node $\psi_{t'} = (\xi_{t'}, s_{t'})$ in the tree, we can define upper and lower value function bounds (cf. Eq. 9.2.1 below) via

$$\boldsymbol{v}^-(\psi_{t'}) = V^{\pi(\xi_{t'})}_{\xi_{t'}}(s_{t'}), \qquad \boldsymbol{v}^+(\psi_{t'}) = V^+_{\xi_{t'}}(s_{t'}),$$

where $\pi(\xi_{t'})$ can be any approximately optimal policy for $\xi_{t'}$. Using backwards induction, we can calculate tighter upper $\boldsymbol{q}^+$ and lower bounds $\boldsymbol{q}^-$ for all non-leaf hyperstates by

$$\boldsymbol{q}^+(\psi_t, a_t) = \sum_{\psi_{t+1}} P(\psi_{t+1} \mid \psi_t, a_t) \left[\rho(\psi_t, a_t) + \gamma\, \boldsymbol{v}^+(\psi_{t+1})\right],$$
$$\boldsymbol{q}^-(\psi_t, a_t) = \sum_{\psi_{t+1}} P(\psi_{t+1} \mid \psi_t, a_t) \left[\rho(\psi_t, a_t) + \gamma\, \boldsymbol{v}^-(\psi_{t+1})\right].$$

We can then use the upper bounds to expand the tree (i.e., to select actions in the tree that maximize $\boldsymbol{v}^+$) while the lower bounds can be used to select the final policy.

Sub-optimal branches can be discarded once their upper bounds become lower than the lower bound of some other branch.

Remark 9.2.1 If $\boldsymbol{q}^-(\psi, a) \geq \boldsymbol{q}^+(\psi, a')$ then a' is sub-optimal at ψ.

However, such an algorithm is only possible to implement when the number of possible MDPs and states are finite. We can generalize this to the infinite case by applying *stochastic* branch and bound methods [3, 4]. This involves estimating upper and lower bounds on the values of leaf nodes through Monte Carlo sampling.

9.2.6 Bounds on the Expected Utility

Bounds on the Bayes-expected utility can serve as a guideline when trying to find a good policy. Accordingly, in this and the following section we aim at obtaining respective upper and lower bounds. First, note that given a belief ξ and a policy π, the respective conditional expected utility is defined as follows.

Definition 9.2.1 (*Bayesian value function π for a belief ξ*)

$$V_\xi^\pi(s) \triangleq \mathbb{E}_\xi^\pi(U \mid s)$$

It is easy to see that the Bayes value function of a policy is simply the expected value function under ξ:

$$V_\xi^\pi(s) = \int_{\mathcal{M}} \mathbb{E}_\mu^\pi(U \mid s)\, \mathrm{d}\xi(\mu) = \int_{\mathcal{M}} V_\mu^\pi(s)\, \mathrm{d}\xi(\mu)$$

However, the Bayes-optimal value function is not equal to the expected value function of the optimal policy for each MDP. In fact, the Bayes-value of any policy is a natural lower bound on the Bayes-optimal value function, as the Bayes-optimal policy is the maximum by definition. We can however use the expected optimal value function as an upper bound on the Bayes-optimal value, that is,

$$\begin{aligned} V_\xi^* &\triangleq \sup_\pi \mathbb{E}_\xi^\pi(U) = \sup_\pi \int_{\mathcal{M}} \mathbb{E}_\mu^\pi(U)\, \mathrm{d}\xi(\mu) \\ &\leq \int_{\mathcal{M}} \sup_\pi V_\mu^\pi\, \mathrm{d}\xi(\mu) = \int_{\mathcal{M}} V_\mu^*\, \mathrm{d}\xi(\mu) \triangleq V_\xi^+. \end{aligned}$$

Given the previous development, it is easy to see that the following inequalities always hold, giving us upper and lower bounds on the value function:

$$V_\xi^\pi \leq V_\xi^* \leq V_\xi^+, \qquad \forall \pi \tag{9.2.1}$$

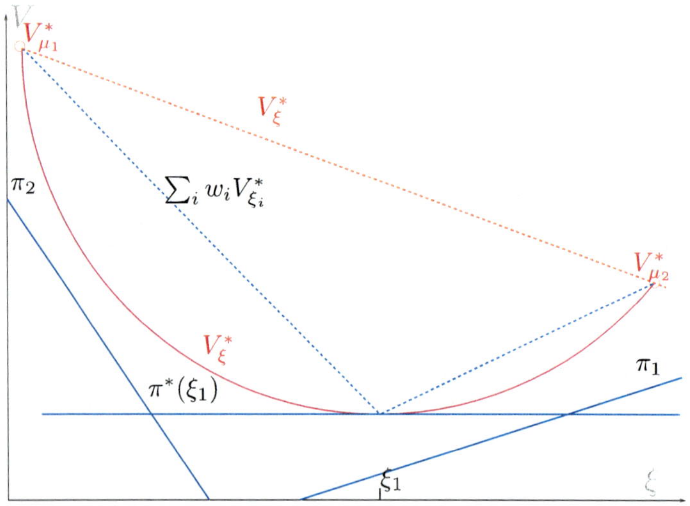

Fig. 9.5 A geometric view of the bounds. Here we plot the expected value of two policies, π_1, π_2 and the policy $\pi^*(\xi_1)$ that is optimal for ξ_1, as well as the Bayes-optimal value function V^*_ξ

These bounds are geometrically demonstrated in Fig. 9.5. They are entirely analogous to the Bayes bounds of Sect. 3.3.1, with the only difference being that we are now considering complete policies rather than simple decisions.

Thompson sampling and upper bounds. The upper bound on the value function, can be easily approximated through Monte Carlo sampling, as shown in Algorithm 9.3 (cf. also the respective policy evaluation Algorithm 9.4). In fact, for $K = 1$, this method is equivalent to *Thompson sampling* [5], which was first used in the context of Bayesian reinforcement learning by [6]. In Thompson sampling, we sample a single MDP from the belief and then act optimally with respect to this, until some exploration condition is met. This is good exploration heuristic with formal performance guarantees for bandit problems [7, 8]. However, obtaining lower bounds requires estimating good policies. An algorithm for doing this through backwards induction will be explained in the following.

Thompson sampling

Algorithm 9.3 Bayesian Monte Carlo policy evaluation

input policy π, belief ξ
for $k = 1, \dots, K$ **do**
  $\mu_k \sim \xi$
  $\boldsymbol{v}_k = V^\pi_{\mu_k}$
end for
$\boldsymbol{u} = \frac{1}{K} \sum_{k=1}^K \boldsymbol{v}_k$
return $\boldsymbol{u}$.

Algorithm 9.4 Bayesian Monte Carlo upper bound

```
input belief ξ
for k = 1, ..., K do
    μ_k ~ ξ
    v_k = V*_{μ_k}
end for
u* = (1/K) Σ_{k=1}^K v_k
return u*
```

9.2.7 *Estimating Lower Bounds on the Value Function with Backwards Induction*

A lower bound on the value function is useful to tell us how tight our upper bounds are. It is possible to obtain one by evaluating any arbitrary policy. So, tighter lower bounds can be obtained by finding better policies, something that was explored by [9, 10].

In particular, we can consider the problem of finding the best memoryless policy. This involves two approximations. Firstly, approximating our belief over MDPs with a sample over a finite set of n MDPs. Secondly, assuming that the belief is nearly constant over time, and performing backwards induction those n MDPs simultaneously. While this greedy procedure might not find the optimal memoryless policy, it still improves the lower bounds considerably (Fig. 9.6).

The central step backwards induction over multiple MDPs is summarized by the following equation, which simply involves calculating the expected utility of a particular policy over all MDPs:

$$Q^{\pi}_{\xi,t}(s,a) \triangleq \int_{\mathcal{M}} \left\{ \bar{r}_{\mu}(s,a) + \gamma \int_{S} V^{\pi}_{\mu,t+1}(s') \, \mathrm{d}P_{\mu}(s' \mid s,a) \right\} \mathrm{d}\xi(\mu) \qquad (9.2.2)$$

The algorithm greedily performs backwards induction as shown in Algorithm 9.5. However, this is not an optimal procedure, since the belief at any time-step t is not constant. Indeed, even though the policy is memoryless, $\xi(\mu \mid s_t, \pi) \neq \xi(\mu \mid s_t, \pi')$. This is because the probability of being at a particular state is different under different policies and at different time-steps (e.g. if you consider periodic MDPs). For the same reason, this type of backwards induction may not converge as value iteration, but can exhibit cyclic convergence similar to the cyclic equilibria in Markov games [11].

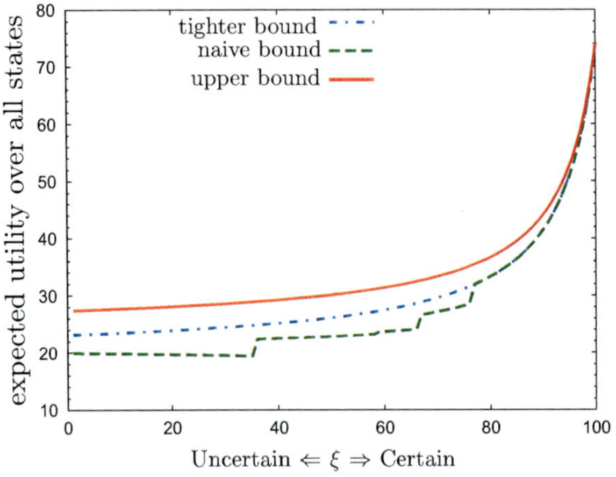

Fig. 9.6 Illustration of the improved bounds. The naive and tighter bound refers to the lower bound obtained by calculating the value of the policy that is optimal for the expected MDP and that obtained by calculating the value of the MMBI policy respectively. The upper bound is V_{ξ}^{+}. The horizontal axis refers to our belief: At the left edge, our belief is uniform over all MDPs, while on the right edge, we are certain about the true MDP

Algorithm 9.5 Multi-MDP backwards induction (MMBI)

input $\mathcal{M}, \xi, \gamma, T$
Set $V^{\pi}_{\mu,T+1}(s) = 0$ for all $s \in \mathcal{S}$.
for $t = T, T-1, \ldots, 0$ **do**
  for $s \in \mathcal{S}, a \in \mathcal{A}$ **do**
    Calculate $Q^{\pi}_{\xi,t}(s, a)$ from (9.2.2) using $\{V^{\pi}_{\mu,t+1}\}$.
  end for
  for $s \in \mathcal{S}$ **do**
    Choose $\pi_t(s) \in \arg\max_{a \in \mathcal{A}} Q_{\xi,t}(s, a)$.
    for $\mu \in \mathcal{M}$ **do**
      Set $V^{\pi}_{\mu,t}(s) = Q^{\pi}_{\mu,t}(s, \pi_t(s))$.
    end for
  end for
end for
return π, Q_{ξ}

In practice, we maintain a belief over an infinite set of MDPs, such as the class of all discrete MDPs with a certain number of state and actions. To apply this algorithm in this case, we can sample a finite number of MDPs from the current belief and then find the optimal policy for this sample, as shown in Algorithm 9.6. For $K = 1$, this is also equivalent to Thompson sampling. However as we can see in Fig. 9.7, Algorithm 9.6 performs better when the number of samples is increased.

Algorithm 9.6 Monte Carlo Bayesian Reinforcement Learning

At time t, sample K MDPs $\mu_1, \ldots, \mu_K$ from ξ_t.
Calculate the best memoryless policy $\pi \approx \arg\max_{\pi \in \Pi_1} \sum_{k=1}^{K} V_{\mu_k}^{\pi}$ wrt the sample.
Execute π until a termination condition is met.

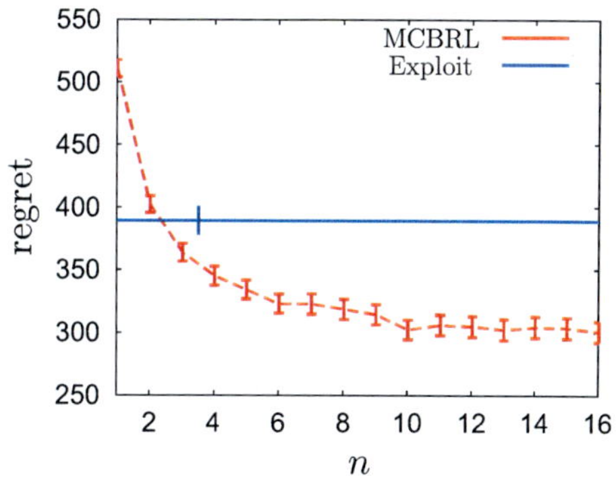

Fig. 9.7 Comparison of the regret between the expected MDP heuristic and sampling with Multi-MDP backwards induction for the chain environment (Example 7.2). The error bars show the standard error of the average regret

9.2.8 *Further Reading*

One of the first treatments of Bayesian reinforcement learning is due to [12]. Although the idea was well-known in the statistical community [13], the first incursion of the idea of Bayes-adaptive policies in reinforcement learning was achieved by Duff's thesis [14]. Most recent advances in Bayes-adaptive policies involve the use of intelligent methods for exploring the tree, such as sparse sampling [31] and Monte Carlo tree search [15].

Instead of sampling MDPs, one could sample beliefs, which leads to a finite hyper-state approximation of the complete belief MDP. One such approach is BEETLE [9, 16], which examines a set of possible future beliefs and approximates the value of each belief with a lower bound. In essence, it then creates the set of policies which are optimal with respect to these bounds.

Another idea is to take advantage of the expectation-maximization view of reinforcement learning [32]. This allows to apply a host of different probabilistic inference algorithms. This approach was investigated by [17].

9.3 Bayesian Methods in Continuous Spaces

Formally, Bayesian reinforcement learning in continuous state spaces is not significantly different from the discrete case. Typically, we assume that the agent acts within a fully observable discrete-time Markov decision process, with a metric state space $\mathcal{S}$, for example $\mathcal{S} \subset \mathbb{R}^d$. The action space $\mathcal{A}$ itself can be either discrete or continuous. The transition kernel can be defined as a collection of probability measures on the continuous state space, indexed by (s, a) as

$$P_\mu(S \mid s, a) \triangleq \mathbb{P}_\mu(s_{t+1} \in S \mid s_t = s, a_t = a), \qquad S \subset \mathcal{S}.$$

There are a number of transition models one can consider for the continuous case. For the purposes of this textbook, we shall limit ourselves to the relatively simple case of linear-Gaussian models.

9.3.1 Linear-Gaussian Transition Models

The simplest type of transition model for an MDP defined on a continuous state space is a linear-Gaussian model, which also results in a closed form posterior calculation due to the conjugate prior. While typically the real system dynamics may not be linear, one can often find some mapping $f : \mathcal{S} \to \mathcal{X}$ to a k-dimensional vector space $\mathcal{X}$ such that the dynamics of the transformed state $\boldsymbol{x}_t \triangleq f(\boldsymbol{s}_t)$ at time t may be well-approximated by a linear system. Then the next state $\boldsymbol{s}_{t+1}$ is given by the output of a function $g : \mathcal{X} \times \mathcal{A} \to \mathcal{S}$ of the transformed state, the action, and some additive noise ε_t, i.e.,

$$\boldsymbol{s}_{t+1} = g(\boldsymbol{x}_t, a_t) + \varepsilon_t.$$

When g is linear and ε_t is normally distributed, this corresponds to a multivariate linear-Gaussian model. In particular, we can parametrize g with a set of $k \times k$-*design matrices* $\{\boldsymbol{A}_i \mid i \in \mathcal{A}\}$, such that $g(\boldsymbol{x}_t, a_t) = \boldsymbol{A}_{a_t}\boldsymbol{x}_t$. We can also define a set of *covariance* matrices $\{\boldsymbol{V}_i \mid i \in \mathcal{A}\}$ for the noise distribution and define the next state distribution by

design matrices

covariance

$$\boldsymbol{s}_{t+1} \mid \boldsymbol{x}_t = \boldsymbol{x}, a_t = i \sim \mathcal{N}(\boldsymbol{A}_i\boldsymbol{x}, \boldsymbol{V}_i).$$

That is, the next state is drawn from a normal distribution with mean $\boldsymbol{A}_i\boldsymbol{x}$ and covariance matrix $\boldsymbol{V}_i$.

In order to model our uncertainty with a (subjective) prior distribution ξ, we have to specify the model structure. Fortunately, in this particular case, a conjugate prior exists in the form of the *matrix-normal distribution* for $\boldsymbol{A}$ and the *inverse-Wishart* distribution for $\boldsymbol{V}$. Given $\boldsymbol{V}_i$, the distribution for $\boldsymbol{A}_i$ is matrix-normal, while the marginal distribution of $\boldsymbol{V}_i$ is inverse-Wishart. More specifically, the dependencies are as follows:

$$\boldsymbol{A}_i \mid \boldsymbol{V}_i = \widehat{\boldsymbol{V}} \sim \phi(\boldsymbol{A}_i \mid \boldsymbol{M}, \boldsymbol{C}, \widehat{\boldsymbol{V}}), \tag{9.3.1}$$

$$\boldsymbol{V}_i \sim \psi(\boldsymbol{V}_i \mid \boldsymbol{W}, n), \tag{9.3.2}$$

where ϕ_i is the prior distribution on dynamics matrices conditional on the covariance and two prior parameters: $\boldsymbol{M}$, which is the prior mean and $\boldsymbol{C}$ which is the prior output (dependent variable) covariance. Finally, ψ is the marginal prior on covariance matrices, which has an inverse-Wishart distribution with $\boldsymbol{W}$ and n. The analytical form of the distributions is given by:

$$\phi(\boldsymbol{A}_i \mid \boldsymbol{M}, \boldsymbol{C}, \widehat{\boldsymbol{V}}) \propto e^{-\frac{1}{2}\operatorname{trace}\left[\boldsymbol{P}(\boldsymbol{A}_i-\boldsymbol{M})\boldsymbol{V}_i^{-1}(\boldsymbol{A}_i-\boldsymbol{M})\boldsymbol{C}\right]}$$

$$\psi(\boldsymbol{V}_i \mid \boldsymbol{W}, n) \propto |\frac{1}{2}\boldsymbol{V}^{-1}\boldsymbol{W}|^{\frac{n}{2}} e^{-\frac{1}{2}\operatorname{trace}(\boldsymbol{V}^{-1}\boldsymbol{W})}$$

Essentially, the considered setting is an extension of the univariate Bayesian linear regression model (see for example [13]) to the multivariate case via vectorization of the mean matrix. Since the prior is conjugate, it is relatively simple to calculate posterior values of the parameters after each observation. While we omit the details, a full description of inference using this model is given by [18].

Further reading. More complex transition models include the non-parametric extension of the above model, namely *Gaussian processes* (GP) [33]. For an n-dimensional state space, one typically applies independent GPs for predicting each state coordinate, i.e., $f_i : \mathbb{R}^n \to \mathbb{R}$. As this completely decouples the state dimensions, it is best to consider a joint model, but this requires various approximations (cf. e.g., [19]). A well-known method for model-based Gaussian process reinforcement learning is GP-Rmax of [34], which has been recently shown by [20] to be KWIK-learnable.[2] *Gaussian processes*

Another straightforward extension of linear models are piecewise linear models, which can be described in a Bayesian non-parametric framework [21]. This avoids the computational complexity that is introduced when using GPs.

9.3.2 Approximate Dynamic Programming

Bayesian methods are also frequently used as part of a dynamic programming approach. Typically, this requires maintaining a distribution over value functions in some sense. For continuous state spaces particularly, one can e.g. assume that the value function $\boldsymbol{v}$ is drawn from a Gaussian process. However, to perform inference we also need to specify some generative model for the observations.

Temporal differences. Engel et al. [22] consider temporal differences from a Bayesian perspective in conjunction with a GP model, so that the rewards are distributed as

$$r_t \mid \boldsymbol{v}, s_t, s_{t+1} \sim \mathcal{N}\big(\boldsymbol{v}(s_{t+1}) - \gamma\, \boldsymbol{v}(s_t), \sigma\big)$$

for any time step t. This essentially gives a simple model for sequences of rewards and states $P(r^T|\boldsymbol{v}, s^T)$. We can now write the posterior as $\xi(\boldsymbol{v} \mid r^T, s^T) \propto P(r^T \mid \boldsymbol{v}, s^T)\,\xi(\boldsymbol{v})$, where the dependence $\xi(\boldsymbol{v}|s^T)$ is suppressed. This model was later updated by [23] using the reward distribution

$$r_t \mid \boldsymbol{v}, s_t, s_{t+1} \sim \mathcal{N}\big(\boldsymbol{v}(s_t) - \gamma\, \boldsymbol{v}(s_{t+1}), N(s_t, s_{t+1})\big),$$

[2] Informally, a class is KWIK-learnable if the number of mistakes made by the algorithm is polynomially bounded in the problem parameters. In the context of reinforcement learning this would be the number of steps for which no guarantee of utility can be provided.

where $N(s, s') \triangleq \Delta_U(s) - \gamma \Delta_U(s')$ with $\Delta_U(s) \triangleq U(s) - \boldsymbol{v}(s)$ denoting the distribution of the residual, i.e., the utility when starting from s minus its expectation. The correlation between $U(s)$ and $U(s')$ is captured via N, and the residuals are modelled as a Gaussian process. While the model is still an approximation, it is equivalent to performing GP regression using Monte Carlo samples of the discounted return.

Bayesian finite-horizon dynamic programming for deterministic systems. Instead of using an approximate model, [24] employ a series of GPs, each for one dynamic programming stage, under the assumption that the dynamics are deterministic and the rewards are Gaussian-distributed. It is possible to extend this approach to the case of non-deterministic transitions, at the cost of requiring additional approximations. However, since a lot of real-world problems do in fact have deterministic dynamics, the approach is consistent.

Bayesian least-squares temporal differences. Tziortziotis and Dimitrakakis [25] instead consider a model for the value function itself, where the random quantity is the empirical transition matrix $\hat{P}$ rather than the reward (which can be assumed to be known):

$$\hat{P}\boldsymbol{v} \mid \boldsymbol{v}, P \sim \mathcal{N}(P\boldsymbol{v}, \beta I)$$

This model makes a different trade-off in its distributional assumptions. It allows us to model the uncertainty about P in a Bayesian manner, but instead of explicitly modelling this as a distribution on P itself, we are modelling a distribution on the resulting Bellman operator.

Gradient methods. As we saw in Sect. 9.2.3, if we are able to sample from the posterior distribution, we can leverage stochastic gradient descent methods to extend any gradient algorithm for reinforcement learning with a given model to the Bayesian setting. In the continuous MDP setting [26] used Gaussian Process models to sample from the posterior. However, other methods could be used, including neural networks or bootstrapping.

9.4 Partially Observable Markov Decision Processes

In most real world applications the state s_t of the system at time t cannot be observed directly. Instead, we obtain some observation x_t, which depends on the state of the system. While this does give us some information about the system state, it is in general not sufficient to pinpoint it exactly. This idea can be formalized as a partially observable Markov decision process (POMDP).

Definition 9.4.1 (*POMDP*) A partially observable Markov decision process (POMDP) μ is a tuple $(\mathcal{X}, \mathcal{S}, \mathcal{A}, P, y)$ where $\mathcal{X}$ is an observation space, $\mathcal{S}$ is a state space, $\mathcal{A}$ is an action space, P is a conditional distribution on observations, states and rewards and y is a starting state distribution. The reward, observation and next state are Markov with respect to the current state and action:

$$\mathbb{P}_\mu(s_{t+1}, r_t, x_t \mid s_t, a_t, \ldots) = P(s_{t+1} \mid s_t, a_t)P(x_t \mid s_t)P(r_t \mid s_t)$$

Here $P(s_{t+1} \mid s_t, a_t)$ is the *transition distribution*, giving the probabilities of next states given the current state and action. $P(x_t \mid s_t)$ is the *observation distribution*, giving the probabilities of different observations given the current state. Finally, $P(r_t \mid s_t)$ is the *reward distribution*, which we make dependent only on the current state for simplicity.

transition distribution
observation distribution
reward distribution

Partially observable Markov decision process
The following graphical model illustrates the dependencies in a POMDP.

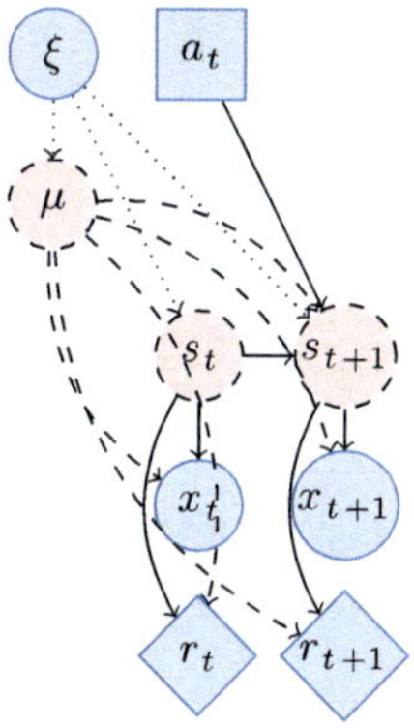

- The system *state* $s_t \in \mathcal{S}$ is not observed.
- We receive an *observation* $x_t \in \mathcal{X}$ and a *reward* $r_t \in \mathcal{R}$.
- We take *action* $a_t \in \mathcal{A}$.
- The system transits to state s_{t+1}.

9.4.1 Solving Known POMDPs

When we know a POMDP's parameters, that is to say, when we know the transition, observation and reward distributions, the problem is formally the same as solving an unknown MDP. In particular, we can similarly define a *belief state* summarizing our knowledge. This takes the form of a probability distribution on the hidden state variable s_t rather than on the model μ. If μ defines starting state probabilities, then the belief is not subjective, as it only relies on the actual POMDP parameters. The transition distribution on states given our belief is as follows.

Belief ξ

For any distribution ξ on $\mathcal{S}$, we define

$$\xi(s_{t+1} \mid a_t, \mu) \triangleq \int_{\mathcal{S}} P_\mu(s_{t+1} \mid s_t a_t) \, \mathrm{d}\xi(s_t).$$

When there is no ambiguity, we shall use ξ to denote arbitrary marginal distributions on states and state sequences given the belief ξ.

When the model μ is given, calculating a belief update is not particularly difficult, but we must take care to properly use the time index t. Starting from Bayes' theorem, it is easy to derive the belief update from ξ_t to ξ_{t+1} as follows.

Belief update

$$\begin{aligned}
\xi_{t+1}(s_{t+1} \mid \mu) &\triangleq \xi_t(s_{t+1} \mid x_{t+1}, r_{t+1}, a_t, \mu) \\
&= \frac{P_\mu(x_{t+1}, r_{t+1} \mid s_{t+1}) \xi_t(s_{t+1} \mid a_t, \mu)}{\xi_t(x_{t+1} \mid a_t, \mu)} \\
\xi_t(s_{t+1} \mid a_t, \mu) &= \int_{\mathcal{S}} P_\mu(s_{t+1} \mid s_t, a_t, \mu) \, \mathrm{d}\xi_t(s_t) \\
\xi_t(x_{t+1} \mid a_t, \mu) &= \int_{\mathcal{S}} P_\mu(x_{t+1} \mid s_{t+1}) \, \mathrm{d}\xi_t(s_{t+1} \mid a_t, \mu)
\end{aligned}$$

A particularly attractive setting is when the model is finite. Then the sufficient statistic also has finite dimension and all updates are in closed form.

Remark 9.4.1 If $\mathcal{S}, \mathcal{A}, \mathcal{X}$ are finite, then we can define a sequence of vectors $\boldsymbol{p}_t \in \mathbb{\Delta}^{|\mathcal{S}|}$ and matrices $\boldsymbol{A}_t$ as

$$\begin{aligned}
\boldsymbol{p}_t(j) &= P(x_t \mid s_t = j), \\
\boldsymbol{A}_t(i, j) &= P(s_{t+1} = j \mid s_t = i, a_t).
\end{aligned}$$

Then writing $\boldsymbol{b}_t(i)$ for $\xi_t(s_t = i)$, we can then use Bayes theorem to obtain

$$\boldsymbol{b}_{t+1} = \frac{\operatorname{diag}(\boldsymbol{p}_{t+1}) \boldsymbol{A}_t \boldsymbol{b}_t}{\boldsymbol{p}_{t+1}^\top \boldsymbol{A}_t \boldsymbol{b}_t}.$$

9.4.2 Solving Unknown POMDPs

Solving a POMDP that is unknown is a much harder problem. The basic update equation for a joint belief on both possible state and possible model is given by

$$\xi(\mu, s^t \mid x^t, a^t) \propto P_\mu(x^t \mid s^t) P_\mu(s^t \mid a^t)\, \xi(\mu).$$

Unfortunately, even for the simplest possible case of two possible models μ_1, μ_2 and binary observations, there is no finite-dimensional representation of the belief at time t.

Strategies for solving unknown POMDPs include solving the full Bayesian decision problem, but this requires exponential inference and planning for exact solutions [27]. For this reason, one usually uses approximations.

One very simple approximation involves replacing a POMDP with a *variable order Markov decision process*, for which inference has only logarithmic computational complexity [28]. The variable order model assumes that the observation probabilities can be decomposed in terms of finite-length *contexts*. Of course, the memory complexity is still linear. This approach has been used by [15] in combination with a Monte Carlo planner for online decision making with promising results.

variable order Markov decision process

In general, finding optimal policies in POMDPs is hard even for restricted classes of policies [35]. However, approximations [29] and stochastic methods as well as policy search methods [30, 32] work quite well in practice.

9.5 Relations Between Different Settings

Markov decision processes can be used to model a wide range of different problems. This section informally (and perhaps sometimes inaccurately) describes the relationship between different MDP settings we have dealt with so far.

When the state and action spaces are finite, the optimal policy for both finite horizon and infinite horizon discounted MDPs can be computed in polynomial time using algorithms like backwards induction or policy iteration (see Chap. 6). However, in other cases obtaining the optimal policy is far from trivial. In the reinforcement learning setting, the MDP is not known and must be estimated while acting in it. If the goal is to maximize expected utility under a prior, the problem becomes a BAMDP. The BAMDP can be seen as a special case of a POMDP, where the underlying latent variable is the MDP parameter, which has a fixed value, rather than the state. For that reason it is generally assumed that BAMDPs have the same complexity as POMDPs. Note however, that POMDPs with linear-Gaussian dynamics can be solved with the exact same controller as linear-Gaussian MDPs, by replacing the actual state with its expected value.

The POMDP problem with discrete state space is similar to a continuous MDP. This is because we can construct an MDP that uses the POMDP belief state as its state. This belief state is finite-dimensional and continuous. If the MDP state space is continuous, then it is in general not possible to decide whether a given policy is optimal in finite time. However, it is possible to check whether a policy is ϵ-optimal under certain regularity conditions on the state space structure. However, if the POMDP state space is discrete, there is only a finite number of possible next belief states for any belief state, and the number of policies is also finite for a finite horizon. Thus, the relationship between these classes is summarized below:

$$\text{d-MDP} \subseteq \text{d-BAMDP} \subseteq \text{d-POMDP} \subseteq \text{c-MDP}$$

While it is known that d-MDP is P-complete and d-POMDP is PSPACE-complete, it is unclear if d-BAMDP is in a simpler class, such as NP.[3]

Finally, the above settings can be generalized to multi-player games. In particular, an MDP with many players and a different reward function for each player is a stochastic game (SG). When the game is zero sum, planning is conjectured to remain in P (although we have not seen a formal proof of this at the time of writing). For non-zero-sum games, computation of a Nash equilibrium has PPAD complexity. Partially observable stochastic games (POSG) are in general in an exponential (or higher) complexity class, even though inference may be simple depending on the type of game.

9.6 Exercises

Exercise 9.1 Consider the algorithms we have seen in Chap. 8. Are any of those applicable to belief-augmented MDPs? Outline a strategy for applying one of those algorithms to the problem. What would be the biggest obstacle we would have to overcome in your specific example?

Exercise 9.2 Prove Remark 9.1.1.

Exercise 9.3 A practical case of Bayesian reinforcement learning in discrete space is when we have an independent belief over the transition probabilities of each state-action pair. Consider the case where we have n states and k actions. Similar to the product-prior in the bandit case in Sect. 6.2, we assign a probability (density) $\xi_{s,a}$ to the probability vector $\boldsymbol{\theta}_{(s,a)} \in \mathbb{\Delta}^n$. We can then define our joint belief on the $(nk) \times n$ matrix $\boldsymbol{\Theta}$ to be

$$\xi(\boldsymbol{\Theta}) = \prod_{s \in \mathcal{S}, a \in \mathcal{A}} \xi_{s,a}(\boldsymbol{\theta}_{(s,a)}).$$

(i) Derive the updates for a product-Dirichlet prior on transitions.
(ii) Derive the updates for a product-Normal-Gamma prior on rewards.
(iii) What would be the meaning of using a Normal-Wishart prior on rewards?

[3] The complexity hierarchy satisfies P $\subseteq$ NP $\subseteq$ PSPACE $\subseteq$ EXP, with P $\subset$ EXP.

Exercise 9.4 Consider the Gaussian process model of Eq. (9.3.2). What is the implicit assumption made about the transition model? If this assumption is satisfied, what does the corresponding posterior distribution represent?

References

1. Csilléry, K., Blum, M.G.B., Gaggiotti, O.E., François, O.: Approximate Bayesian computation (ABC) in practice. Trends Ecol. Evol. **25**(7), 410–418 (2010)
2. Dimitrakakis, C., Tziortziotis, N.: ABC reinforcement learning. In Proceedings of the 30th International Conference on Machine Learning, ICML 2013, pp. 684–692 (2013). (JMLR.org)
3. Dimitrakakis, C.: Complexity of stochastic branch and bound methods for belief tree search in Bayesian reinforcement learning. In: 2nd International Conference on Agents and Artificial Intelligence (ICAART 2010), pp. 259–264. Springer, Valencia, Spain (2010)
4. Dimitrakakis, C.: Tree exploration for Bayesian RL exploration. In: 2008 International Conferences on Computational Intelligence for Modelling, Control and Automation (CIMCA 2008), Intelligent Agents, Web Technologies and Internet Commerce (IAWTIC 2008), Innovation in Software Engineering (ISE 2008), pp. 1029–1034. IEEE Computer Society (2008)
5. Thompson, W.R.: On the likelihood that one unknown probability exceeds another in view of the evidence of two samples. Biometrika **25**(3–4), 285–294 (1933)
6. Strens, M.J.A.: A Bayesian framework for reinforcement learning. In: Proceedings of the Seventeenth International Conference on Machine Learning (ICML 2000), pp. 943–950. Morgan Kaufmann (2000)
7. Kaufmann, E., Korda, N., Munos, R.: Thompson sampling: an asymptotically optimal finite-time analysis. In: Algorithmic Learning Theory-23rd International Conference, ALT 2012. Proceedings. Lecture Notes in Computer Science, vol. 7568, pp. 199–213. Springer (2012)
8. Osband, I., Russo, D., Van Roy, B.: (More) efficient reinforcement learning via posterior sampling. In: Advances in Neural Information Processing Systems, vol. 26, pp. 3003–3011 (2013)
9. Poupart, P., Vlassis, N.A., Hoey, J., Regan, K.: An analytic solution to discrete Bayesian reinforcement learning. In: Machine Learning, Proceedings of the 23rd International Conference (ICML 2006), pp. 697–704. ACM (2006)
10. Dimitrakakis, C.: Robust Bayesian reinforcement learning through tight lower bounds. In: Sanner, S., Hutter, M. (eds.) Recent Advances in Reinforcement Learning–9th European Workshop, EWRL 2011. Lecture Notes in Computer Science, vol. 7188, pp. 177–188. Springer (2011)
11. Zinkevich, M., Greenwald, A., Littman, M.L.: Cyclic equilibria in Markov games. In: Advances in Neural Information Processing Systems, vol. 18, pp. 1641–1648 (2006)
12. Richard Ernest Bellman: A problem in the sequential design of experiments. Sankhya **16**, 221–229 (1957)
13. DeGroot, M.H.: Optimal Statistical Decisions. Wiley (1970)
14. Duff, M.O.: Optimal learning computational procedures for Bayes-adaptive Markov decision processes. Ph.D. thesis, University of Massachusetts at Amherst (2002)
15. Veness, J., Ng, K.S., Hutter, M., Silver, D.: A Monte Carlo AIXI approximation. Technical Report 0909.0801 (2009). (arXiv)
16. Poupart, P., Vlassis, N.: Model-based Bayesian reinforcement learning in partially observable domains. In: International Symposium on Artificial Intelligence and Mathematics, ISAIM 2008 (2008)
17. Furmston, T, Barber, D.: Variational methods for reinforcement learning. In: Proceedings of the 13th International Conference on Artificial Intelligence and Statistics (AISTATS 2010), pp. 241–248 (2010)
18. Minka, T.P.: Bayesian linear regression. Technical Report, Microsoft research (2000)

19. Álvarez, M., Luengo, D., Titsias, M., Lawrence, N.: Efficient multioutput Gaussian processes through variational inducing kernels. In: Proceedings of the 13th International Conference on Artificial Intelligence and Statistics (AISTATS 2010), pp. 25–32 (2010)
20. Grande, R.C., Walsh, T.J., How, J.P.: Sample efficient reinforcement learning with gaussian processes. In: Proceedings of the 31th International Conference on Machine Learning, ICML 2014, pp. 1332–1340 (2014). (JMLR.org)
21. Tziortziotis, N., Dimitrakakis, C., Blekas, K.: Cover tree Bayesian reinforcement learning. J. Mach. Learn. Res. **15**(1), 2313–2335 (2014)
22. Engel, Y., Mannor, S., Meir, R.: Bayes meets Bellman: the Gaussian process approach to temporal difference learning. In: Machine Learning, Proceedings of the 20th International Conference (ICML 2003), pp. 154–161. AAAI Press (2003)
23. Engel, Y., Mannor, S., Meir, R.: Reinforcement learning with gaussian processes. In: Machine Learning, Proceedings of the 22nd International Conference (ICML 2005), pp. 201–208. ACM (2005)
24. Deisenroth, M.P., Rasmussen, C.E., Peters, J.: Gaussian process dynamic programming. Neurocomputing **72**(7–9), 508–1524 (2009)
25. Tziortziotis, N., Dimitrakakis, C.: Bayesian inference for least squares temporal difference regularization. In: Machine Learning and Knowledge Discovery in Databases-European Conference, ECML PKDD 2017, Proceedings Part II. Lecture Notes in Computer Science, vol. 10535, pp. 126–141. Springer (2017)
26. Ghavamzadeh, M., Engel, Y.: Bayesian policy gradient algorithms. In: Advances in Neural Information Processing Systems, vol. 19, pp. 457–464. MIT Press (2006)
27. Ross, S., Chaib-draa, B., Pineau, J.: Bayes-adaptive POMDPs. In: Advances in Neural Information Processing Systems, vol. 20, pp. 1225–1232 (2008)
28. Dimitrakakis, C.: Bayesian variable order Markov models. In: Proceedings of the 13th International Conference on Artificial Intelligence and Statistics (AISTATS 2010), pp. 161–168 (2010)
29. Spaan, M.T.J., Vlassis, N.: Perseus: randomized point-based value iteration for POMDPs. J. Artif. Intell. Res. **24**(1), 195–220 (2005)
30. Baxter, J., Bartlett, P.L.: Reinforcement learning in POMDP's via direct gradient ascent. In: Proceedings of the 17th International Conference on Machine Learning, ICML 2000, pp. 41–48. Morgan Kaufmann, San Francisco, CA (2000)
31. Wang, T., Lizotte, D., Bowling, M., Schuurmans, D.: Bayesian sparse sampling for on-line reward optimization. In: Machine Learning, Proceedings of the 22nd International Conference (ICML 2005). ACM (2005)
32. Toussaint, M., Harmelign, S., Storkey, A.: Probabilistic inference for solving (PO)MDPs. Technical Report EDI-INF-RR-0934, University of Endinburgh, School of Informatics (2006)
33. Rasmussen, C.E., Williams, C.K.I.: Gaussian Processes for Machine Learning. MIT Press (2006)
34. Jung, T., Stone, P.: Gaussian processes for sample-efficient reinforcement learning with RMAX-like exploration. In: Machine Learning and Knowledge Discovery in Databases, European Conference, ECML PKDD 2010. Lecture Notes in Computer Science, vol. 6321, pp. 601–616. Springer (2010)
35. Vlassis, N., Littman, M.L., Barber, D.: On the computational complexity of stochastic controller optimization in POMDPs. ACM Trans. Comput. Theory **4**(4), 12:1–12:8 (2012)

Chapter 10
Distribution-Free Reinforcement Learning

10.1 Introduction

The Bayesian framework requires specifying a prior distribution. For many reasons, we may frequently be unable to do that. In addition, as we have seen, the Bayes-optimal solution is often intractable. In this chapter we shall take a look at algorithms that do not require specifying a prior distribution. Instead, they employ the heuristic of "optimism under uncertainty" to select policies. This idea is very similar to heuristic search algorithms, such as A^* [1]. All these algorithms assume the best possible model that is consistent with the observations so far and choose the optimal policy in this "optimistic" model. Intuitively, this means that for each possible policy we maintain an upper bound on the value/utility we can reasonably expect from it. In general we want this upper bound to

1. be as tight as possible (i.e., to be close to the true value),
2. still hold with high probability.

We begin with an introduction to these ideas in bandit problems, when the objective is to maximize total reward. We then expand this discussion to structured bandit problems, which have many applications in optimization. Finally, we look at the case of maximizing total reward in unknown MDPs.

10.2 Finite Stochastic Bandit Problems

First of all, let us briefly recall the stochastic bandit setting, which we already have considered in Sect. 6.2. The learner in discrete time steps $t = 1, 2, \dots$ chooses an *arm* a_t from a given set $\mathcal{A} = \{1, \dots, K\}$ of K arms. The rewards r_t the learner obtains in return are random and assumed to be independent as well as bounded, e.g., $r_t \in [0, 1]$. The expected reward $r(i) = \mathbb{E}(r_t | a_t = i)$ for choosing any arm i

C. Dimitrakakis and R. Ortner, *Decision Making Under Uncertainty and Reinforcement Learning*, Intelligent Systems Reference Library 223,
https://doi.org/10.1007/978-3-031-07614-5_10

is unknown to the learner, who aims to maximize the total reward $\sum_{t=1}^{T} r_t$ after a certain number of T time steps.

Let $r^* \triangleq \max_i r(i)$ be the highest expected reward that can be achieved. Obviously, the optimal policy π^* in each time step chooses the arm giving the highest expected reward r^*. The learner who does not know which arm is optimal will choose at each time step t an arm a_t from $\mathcal{A}$, or more generally, a probability distribution over the arms from which a_t then is drawn. It is important to notice that maximizing the total reward is equivalent to minimizing total regret with respect to that policy.

Definition 10.2.1 (*Total regret*) The *(total) regret* of a policy π relative to the optimal fixed policy π^* after T steps is

$$L_T(\pi) \triangleq \sum_{t=1}^{T} \left(r_t^* - r_t^\pi\right),$$

where r_t^π is the reward obtained by the policy π at step t and $r_t^* \triangleq r_t^{\pi^*}$. Accordingly, the *expected (total) regret* is

$$\mathbb{E}\, L_T(\pi) \triangleq T r^* - \mathbb{E}_\pi \sum_{t=1}^{T} r_t.$$

The regret compares the collected rewards to those of the best fixed policy. Comparing instead to the best rewards obtained by the arms at each time would be too hard due to their randomness.

We note that the notion of regret we consider here is usually called *pseudo-regret*, while the term *expected regret* often refers to the just mentioned comparison to the actual best rewards, cf. [8] for a more detailed discussion.

10.2.1 The UCB1 Algorithm

It makes sense for a learning algorithm to use the empirical average rewards obtained for each arm so far.

Empirical average

$$\hat{r}_{t,i} \triangleq \frac{1}{N_{t,i}} \sum_{k=1}^{t} r_{k,i}\, \mathbb{I}\{a_k = i\}, \qquad \text{where } N_{t,i} \triangleq \sum_{k=1}^{t} \mathbb{I}\{a_k = i\}$$

and $r_{k,i}$ denotes the (random) reward the learner receives upon choosing arm i at step k.

Simply always choosing the arm with best the empirical average reward so far is not the best idea, because you might get stuck with a sub-optimal arm: If the optimal arm underperforms at the beginning, so that its empirical average is far below the true mean of a suboptimal arm, it will never be chosen again. A better strategy is to choose arms optimistically. Intuitively, as long as an arm has a significant chance of being the best, you play it every now and then. One simple way to implement this is shown in the following UCB1 algorithm Auer et al. [2].

Algorithm 10.1 UCB1

input $\mathcal{A}$
Choose each arm once to obtain an initial estimate.
for $t = 1, \ldots$ **do**
  Choose arm $a_t = \arg\max_{i\in\mathcal{A}} \left\{\hat{r}_{t-1,i} + \sqrt{\frac{2\ln t}{N_{t-1,i}}}\right\}$.
end for

Thus, the algorithm adds a bonus value of order $O(\sqrt{\ln t/N_{t,i}})$ to the empirical value of each arm thus forming an *upper confidence bound*. This upper confidence bound value is such that the true mean reward of each arm will lie below it with high probability by the Hoeffding bound (4.5.5).

upper confidence bound

Theorem 10.2.1 (Auer et al. [2]) *The expected regret of UCB1 after T time steps is at most*

$$\mathbb{E}\, L_T(UCB1) \leq \sum_{i:r(i)<r^*} \frac{8\ln T}{r^* - r(i)} + 5\sum_i (r^* - r(i)).$$

Proof By Wald's identity (5.10) the expected regret can be written as

$$\mathbb{E}\, L_T = \mathbb{E}\sum_{t=1}^{T}(r^* - r_t) = \sum_i (r^* - r(i))\,\mathbb{E}\, N_{T,i}, \tag{10.2.1}$$

so that we focus on bounding $\mathbb{E}\, N_{t,i}$. Thus, let i be an arbitrary suboptimal arm, for which we shall consider when it will be chosen by the algorithm. Write $B_{t,s} = \sqrt{(2\ln t)/s}$ for the bonus value at step t after s observations. Note that for fixed values of $t, s, s_i \in \mathbb{N}$ under the assumption that $N_{t,i} = s_i$ and (the count of the optimal action) $N_{t,*} = s$, we have by the Hoeffding bound (4.5.5) that

$$\mathbb{P}(\hat{r}_{t,i} \geq r(i) + B_{t,s_i}) \leq e^{-4\ln t} = t^{-4},$$
$$\mathbb{P}(\hat{r}_{t,*} \leq r^* - B_{t,s}) \leq e^{-4\ln t} = t^{-4}.$$

Accordingly we may assume that (taking care of the contribution of the error probabilities to $\mathbb{E}\, N_{t,i}$ below)

$$\hat{r}_{t,i} < r(i) + B_{t,N_{t,i}}, \tag{10.2.2}$$
$$r^* < \hat{r}_{t,*} + B_{t,N_{t,*}}. \tag{10.2.3}$$

Now note that for $s \geq \lceil (8 \ln T)/(r^* - r(i))^2 \rceil$ it holds that

$$2B_{t,s} \leq (r^* - r(i)), \tag{10.2.4}$$

so that after arm i has been chosen $\lceil (8 \ln T)/(r^* - r(i))^2 \rceil$ times we get from (10.2.2), (10.2.4), and (10.2.3) that

$$\begin{aligned} \hat{r}_{t,i} + B_{t,N_{t,i}} &< r(i) + 2B_{t,N_{t,i}} \leq r^* \\ &< \hat{r}_{t,*} + B_{t,N_{t,*}}. \end{aligned}$$

This shows that after $\lceil (8 \ln T)/(r^* - r(i))^2 \rceil$ samples from arm i, the algorithm won't choose it again. Taking into account the error probabilities for (10.2.2) and (10.2.3), arm i may be played once at each step t whenever either equation does not hold. Summing over all possible values for t, $N_{t,i}$ and $N_{t,*}$ this shows that

$$\mathbb{E} N_{t,i} \leq \left\lceil \frac{8 \ln T}{(r^* - r(i))^2} \right\rceil + \sum_{\tau \geq 1} \sum_{s \leq \tau} \sum_{s_i \leq \tau} 2\tau^{-4}.$$

Combining this with (10.2.1) and noting that the sum converges to a value < 4, proves the regret bound. □

The UCB1 algorithm is actually not the first algorithm employing *optimism in the face of uncertainty* to deal with the exploration-exploitation dilemma, nor the first that uses confidence intervals for that purpose. This idea goes back to the seminal work of Lai and Robbins [3] that used the same approach, however in a more complicated form. In particular, the whole history is used for computing the arm to choose. The derived bounds of Lai and Robbins [3] show that after T steps each suboptimal arm is played at most $\left(\frac{1}{D_{\mathrm{KL}}} + o(1)\right) \log T$ times in expectation, where D_{KL} measures the distance between the reward distributions of the optimal and the suboptimal arm by the Kullback-Leibler divergence, and $o(1) \to 0$ as $T \to \infty$. This bound was also shown to be asymptotically optimal Lai and Robbins [3]. A lower bound logarithmic in T for any finite T that is close to matching the bound of Theorem 10.2.1 can be found in Mannor and Tsitsiklis [4]. Improvements that get closer to the lower bound (and are still based on the UCB1 idea) can be found in Auer and Ortner [5], while the gap has been finally closed by Lattimore [6].

For so-called distribution-independent bounds that do not depend on problem parameters like the "gaps" $r^* - r(i)$, see e.g. Audibert and Bubeck [7]. In general, these bounds cannot be logarithmic in T anymore, as the gaps may be of order $1/\sqrt{T}$ resulting in bounds that are $O(\sqrt{KT})$, just like in the nonstochastic setting that we will take a look at next.

10.2.2 Non i.i.d. Rewards

The stochastic setting just considered is only one among several variants of the multi-armed bandit problem. While it is impossible to cover them all, we give a brief of the most common scenarios and refer to Bubeck and Cesa-Bianchi [8] for a more complete overview.

What is common to most variants of the classic stochastic setting is that the assumption of receiving i.i.d. rewards when sampling a fixed arm is loosened. The most extreme case is the so-called *nonstochastic*, sometimes also termed *adversarial bandit* setting, where the reward sequence for each arm is assumed to be fixed in advance (and thus not random at all). In this case, the reward is maximized when choosing in each time step the arm that maximizes the reward at this step. Obviously, since the reward sequences can be completely arbitrary, no learner can stand a chance to perform well with respect to this optimal policy. Thus, one confines oneself to consider the regret with respect to the best *fixed* arm in hindsight, that is, $\arg\max_i \sum_{t=1}^{T} r_{t,i}$ where $r_{t,i}$ is the reward of arm i at step t. It is still not clear that this is not too much to ask for, but it turns out that one can achieve regret bounds of order $O(\sqrt{KT})$ in this setting. Clearly, algorithms that choose arms deterministically can always be tricked by an adversarial reward sequence. However, algorithms that at each time step choose an arm from a suitable distribution over the arms (that is updated according to the collected rewards), can be shown to give the mentioned optimal regret bound. A prominent exponent of these algorithms is the Exp3 algorithm of Auer et al. [9], that uses an exponential weighting scheme. *adversarial bandit*

In the *contextual bandit* setting the learner receives some additional side information called the *context*. The reward for choosing an arm is assumed to depend on the context as well as on the chosen arm and can be either stochastic or adversarial. The learner usually competes against the best policy that maps contexts to arms. There is a notable amount of literature dealing with various settings that are usually also interesting for applications like web advertisement where user data takes the role of provided side information. For an overview see e.g. Chap. 4 of Bubeck and Cesa-Bianchi [8] or Part V of Lattimore and Szepesvári [10]. *contextual bandit*

In other settings the i.i.d. assumption about the rewards of a fixed arm is replaced by more general assumptions, such as that underlying each arm there is a Markov chain and rewards depend on the state of the Markov chain when sampling the arm. This is called the *restless bandits* problem, that is already quite close to the general reinforcement learning setting with an underlying Markov decision process (see Sect. 10.3.1). Regret bounds in this setting can be shown to be $\tilde{O}(\sqrt{T})$ even if at each time step the learner can observe only the state of the arm he chooses, see Ortner et al. [11].

10.3 Reinforcement Learning in MDPs

Taking a step further from the bandit problems of the previous sections we now want to consider a more general reinforcement learning setting where the learner operates on an unknown underlying MDP. Note that the stochastic bandit problem corresponds to a single state MDP.

Thus, consider an MDP μ with state space $\mathcal{S}$, action space $\mathcal{A}$, and let $r(s, a) \in [0, 1]$ and $P(\cdot|s, a)$ be the mean reward and the transition probability distribution on $\mathcal{S}$ for each state $s \in \mathcal{S}$ and each action $a \in \mathcal{A}$, respectively. For the moment we assume that $\mathcal{S}$ and $\mathcal{A}$ are finite. As we have seen in Sect. 6.6 there are various optimality criteria for MDPs. In the spirit of the bandit problems considered so far we consider undiscounted rewards and examine the regret after any T steps with respect to an optimal policy.

Since the optimal T-step policy in general will be non-stationary and different for different horizons T and different initial states, we will compare to a *gain optimal* policy π^* as introduced in Definition 6.6.3. Further, we assume that the MDP is *communicating*. That is, for any two states s, s' there is a policy $\pi_{s,s'}$ that with positive probability reaches s' when starting in s. This assumption allows the learner to recover when making a mistake. Note that in MDPs that are not communicating one wrong step may lead to a suboptimal region of the state space that cannot be left anymore, which makes competing to an optimal policy in a learning setting impossible. For communicating MDPs we can define the *diameter* to be the maximal expected time it takes to connect any two states.

Definition 10.3.1 Let $T(\pi, s, s')$ the expected number of steps it takes to reach state s' when starting in s and playing policy π. Then the *diameter* is defined as

$$D \triangleq \max_{s,s'} \min_{\pi} T(\pi, s, s').$$

Given that our rewards are assumed to be bounded in $[0, 1]$, intuitively, when we make one wrong step in some state s, in the long run we won't lose more than D. After all, in D steps we can go back to s and continue optimally.

Under the assumption that the MDP is communicating, the gain g^* can be shown to be independent of the initial state, that is, $g^*(s) = g^*$ for all states s. Accordingly, we define the T-step regret of a learning algorithm as

$$L_T \triangleq \sum_{t=1}^{T} (g^* - r_t),$$

where r_t is the reward collected by the algorithm at step t. Note that in general (and depending on the initial state) the value Tg^* we compare to will differ from the optimal T-step reward. However, this difference can be shown to be upper bounded by the diameter and is therefore negligible when considering the regret.

10.3.1 An Upper-Confidence Bound Algorithm

Now we aim at extending the idea underlying the UCB1 algorithm to the general reinforcement learning setting. Again, we would like to have for each (stationary) policy π an upper bound on the gain that is reasonable to expect. Note that simply taking each policy to be the arm of a bandit problem does not work well. First, to approach the true gain of a chosen policy, it will not be sufficient to choose it just once. It would be necessary to follow each policy for a sufficiently high number of consecutive steps. Without knowledge of some characteristics of the underlying MDP like mixing times, it might be however difficult to determine how long a policy shall be played. Further, due to the large number of stationary policies, which is $|\mathcal{A}|^{|\mathcal{S}|}$, the regret bounds that would result from such an approach would be exponential in the number of states.

Thus, we rather maintain confidence regions for the rewards and transition probabilities of each state-action pair s, a. Then, at each step t, these confidence regions implicitly also define a confidence region for the true underlying MDP μ^*, that is, a set M_t of *plausible* MDPs. For suitably chosen confidence intervals for the rewards and transition probabilities one can obtain that

$$\mathbb{P}(\mu^* \notin M_t) < \delta. \tag{10.3.1}$$

Given this confidence region M_t, one can define the optimistic value for any policy π to be

$$g_+^\pi(M_t) \triangleq \max \left\{ g_\mu^\pi \;\middle|\; \mu \in M_t \right\}. \tag{10.3.2}$$

Note that similar to the bandit setting this estimate is optimistic for each policy, as due to (10.3.1) it holds that $g_+^\pi(M_t) \geq g_\mu^\pi$ with high probability. Analogously to UCB1 we would like to make an optimistic choice among the possible policies, that is, we choose a policy π that maximizes $g_+^\pi(M_t)$.

However, unlike in the bandit setting where we immediately receive a sample from the reward of the chosen arm, in the MDP setting we only obtain information about the reward in the current state. Thus, we should not play the chosen optimistic policy just for one but a sufficiently large number of steps. An easy way is to play policies in episodes of increasing length, such that sooner or later each action is played for a sufficient number of steps in each state. Summarized, we obtain (the outline of) an algorithm as shown below.

UCRL2 [12] outline

In episodes $k = 1, 2, \ldots$

- At the first step t_k of episode k, update the confidence region M_{t_k}.
- Compute an optimistic policy $\tilde{\pi}_k \in \arg\max_\pi g_+^\pi(M_{t_k})$.
- Execute $\tilde{\pi}_k$, observe rewards and transitions until t_{k+1}.

10.3.1.1 Technical Details for UCRL2

To make the algorithm complete, we have to fill in some technical details. In the following, let S be the number of states and A the number of actions of the underlying MDP μ. Further, the algorithm takes a confidence parameter $\delta > 0$.

The confidence region. Concerning the confidence regions, for the rewards it is sufficient to use confidence intervals similar to those for UCB1. For the transition probabilities we consider all those transition probability distributions to be plausible whose $\|\cdot\|_1$-norm is close to the empirical distribution $\hat{\boldsymbol{P}}_t(\cdot \mid s, a)$. That is, the confidence region M_t at step t used to compute the optimistic policy can be defined as the set of MDPs with mean rewards $r(s, a)$ and transition probabilities $\boldsymbol{P}(\cdot \mid s, a)$ such that

$$\left|r(s,a) - \hat{r}(s,a)\right| \le \sqrt{\tfrac{7\log(2SAt/\delta)}{2N_t(s,a)}},$$

$$\left\|\boldsymbol{P}(\cdot \mid s,a) - \hat{\boldsymbol{P}}_t(\cdot \mid s,a)\right\|_1 \le \sqrt{\tfrac{14S\log(2At/\delta)}{N_t(s,a)}},$$

where $\hat{r}(s, a)$ and $\hat{\boldsymbol{P}}_t(\cdot \mid s, a)$ are the estimates for the rewards and the transition probabilities, and $N_t(s, a)$ denotes the number of samples of action a in state s at time step t.

One can show via a bound due to Weissman et al. [13] that given n samples of the transition probability distribution $\boldsymbol{P}(\cdot \mid s, a)$, one has

$$\mathbb{P}\left(\left\|\boldsymbol{P}(\cdot \mid s,a) - \hat{\boldsymbol{P}}_t(\cdot \mid s,a)\right\|_1 \ge \epsilon\right) \le 2^S \exp\left(-\frac{n\epsilon}{2}\right).$$

Using this together with standard Hoeffding bounds for the reward estimates, it can be shown that the confidence region contains the true underlying MDP with high probability.

Lemma 10.3.1

$$\mathbb{P}(\mu^* \in M_t) > 1 - \frac{\delta}{15t^6}.$$

Episode lengths. Concerning the termination of episodes, as already mentioned, we would like to have episodes that are long enough so that we do not suffer large regret when playing a suboptimal policy. Intuitively, it only pays off to recompute the optimistic policy when the estimates or confidence intervals have changed sufficiently. One option is e.g. to terminate an episode when the confidence interval for one state-action pair has shrinked by some factor. Even simpler, one can terminate an episode when a state-action pair has been sampled often (compared to the samples one had before the episode has started), e.g. when one has doubled the number of visits in some state-action pair. This also allows to bound the total number of episodes up to step T.

Lemma 10.3.2 *If an episode of UCRL2 is terminated when the number of visits in some state-action pair has been doubled, the total number of episodes up to step T is upper bounded by $SA \log_2 \frac{8T}{SA}$.*

This episode termination criterion also allows to bound the sum over all fractions of the form $\frac{v_k(s,a)}{\sqrt{N_k(s,a)}}$, where $v_k(s, a)$ is the number of times action a has been chosen *in* state s during episode k, while $N_k(s, a)$ is the respective count of visits *before* episode k. The evaluation of this sum will turn out to be important in the regret analysis below to bound the sum over all confidence intervals over the visited state-action pairs.

Lemma 10.3.3

$$\sum_k \sum_{s,a} \frac{v_k(s, a)}{\sqrt{N_k(s, a)}} \leq (\sqrt{2} + 1)\sqrt{SAT}.$$

Calculating the optimistic policy. It is important to note that the computation of the optimistic policy can be performed efficiently by using a modification of value iteration. Intuitively, for each policy π the optimistic value $g_+^\pi(M_t)$ maximizes the gain over all possible values in the confidence intervals for the rewards and the transition probabilities for π. This is an optimization problem over a compact space that can be easily solved. More precisely, in order to find $\arg\max_\pi g_+^\pi(M_t)$, for each considered policy one additionally has to determine the precise values for rewards and transition probabilities within the confidence region. This corresponds to finding the optimal policy in an MDP with compact action space, which can be solved by an extension of value iteration that in each iteration now not only maximizes over the original action space but also within the confidence region of the respective action. Noting that $g_+^\pi(M_t)$ is maximized when the rewards are set to their upper confidence values, this results in the following value iteration scheme:

1. Set the optimistic rewards $\tilde{r}(s, a)$ to the upper confidence values for all states s and all actions a.
2. Set $u_0(s) := 0$ for all s.
3. For $i = 0, 1, 2, \ldots$ set

$$u_{i+1}(s) := \max_a \left\{ \tilde{r}(s, a) + \max_{P \in \mathcal{P}(s,a)} \left\{ \sum_{s'} P(s')\, u_i(s') \right\} \right\}, \qquad (10.3.3)$$

where $\mathcal{P}(s, a)$ is the set of all plausible transition probabilities for choosing action a in state s.

Similarly to the value iteration algorithm in Sect. 6.5.4.1, this scheme can be shown to converge. More precisely, one can show that $\max_s\{u_{i+1}(s) - u_i(s)\} - \min_s\{u_{i+1}(s) - u_i(s)\} \to 0$ and also

$$u_{i+1}(s) \to u_i(s) + g_+^{\tilde{\pi}} \text{ for all } s. \qquad (10.3.4)$$

After convergence the maximizing actions constitute the optimistic policy $\tilde{\pi}$, and the maximizing transition probabilities are the respective optimistic transition values $\tilde{P}$.

One can also show that the so-called *span* $\max_s u_i(s) - \min_s u_i(s)$ of the converged value vector u_i is upper bounded by the diameter. This follows by optimality of the vector u_i. Intuitively, if the span would be larger than D one could increase the collected reward in the lower value state s^- by going (as fast as possible) to the higher value state s^+. Note that this argument uses the fact that the true MDP is plausible w.h.p., so that we may take the true transitions to get from s^- to s^+.

Lemma 10.3.4 *Let $u_i(s)$ the converged value vector. Then*

$$\max_s u_i(s) - \min_s u_i(s) \leq D.$$

10.3.1.2 Analysis of UCRL2

In this section we derive the following regret bound for UCRL2.

Theorem 10.3.1 ([12]) *In an MDP with S states, A actions, and diameter D with probability of at least $1 - \delta$ the regret of UCRL2 after any T steps is bounded by*

$$\text{const} \cdot DS\sqrt{AT \log\left(\tfrac{T}{\delta}\right)}.$$

Proof The main idea of the proof is that by Lemma 10.3.1 we have that

$$\tilde{g}_k^* \triangleq g_+^{\tilde{\pi}_k}(M_{t_k}) \;\geq\; g^* \;\geq\; g^{\tilde{\pi}_k}, \tag{10.3.5}$$

so that the regret in each step is upper bounded by the width of the confidence interval for $g^{\tilde{\pi}_k}$, that is, by $\tilde{g}_k^* - g^{\tilde{\pi}_k}$. In what follows we need to break down this confidence interval to the confidence intervals we have for rewards and transition probabilities.

In the following, we consider that the true MDP μ is always contained in the confidence regions M_t considered by the algorithm. Using Lemma 10.3.1 it is not difficult to show that with probability at least $1 - \frac{\delta}{12T^{5/4}}$ the regret accumulated due to $\mu \notin M_t$ at some step t is bounded by $\sqrt{T}$.

Further, note that the random fluctuation of the rewards can be easily bounded by Hoeffding's inequality (4.5.5), that is, if s_t and a_t denote the state and action at step t, we have

$$\sum_{t=1}^{T} r_t \geq \sum_t r(s_t, a_t) - \sqrt{\tfrac{5}{8}T \log \tfrac{8T}{\delta}}$$

with probability at least $1 - \frac{\delta}{12T^{5/4}}$.

Therefore, writing $v_k(s, a)$ for the number of times action a has been chosen in state s in episode k we have $\sum_t r(s_t, a_t) = \sum_k \sum_{s,a} v_k(s, a)\, r(s, a)$ so that by (10.3.5) we can bound the regret by

$$\sum_{t=1}^{T}(g^* - r_t) \leq \sum_k \sum_{s,a} v_k(s,a)\big(\tilde{g}_k^* - r(s,a)\big) + \sqrt{T} + \sqrt{\tfrac{5}{8}T \log \tfrac{8T}{\delta}} \quad (10.3.6)$$

with probability at least $1 - \frac{2\delta}{12T^{5/4}}$.

Thus, let us consider an arbitrary but fixed episode k, and consider the regret

$$\sum_{s,a} v_k(s,a)\big(\tilde{g}_k^* - r(s,a)\big)$$

the algorithm accumulates in this episode. Let $\mathrm{conf}_k^r(s,a)$ and $\mathrm{conf}_k^p(s,a)$ be the width of the confidence intervals for rewards and transition probabilities in episode k. First, we simply have

$$\begin{aligned}\sum_{s,a} v_k(s,a)\big(\tilde{g}_k^* - r(s,a)\big) &\leq \sum_{s,a} v_k(s,a)\big(\tilde{g}_k^* - \tilde{r}_k(s,a)\big) \\ &\quad + \sum_{s,a} v_k(s,a)\big(\tilde{r}_k(s,a) - r(s,a)\big), \quad (10.3.7)\end{aligned}$$

where the second term is bounded by

$$|\tilde{r}_k(s,a) - \hat{r}_k(s,a)| + |\hat{r}_k(s,a) - r(s,a)| \leq 2\mathrm{conf}_k^r(s,a)$$

w.h.p. by Lemma 10.3.1, so that

$$\sum_{s,a} v_k(s,a)\big(\tilde{r}_k(s,a) - r(s,a)\big) \leq 2\sum_{s,a} v_k(s,a) \cdot \mathrm{conf}_k^r(s,a). \quad (10.3.8)$$

For the first term in (10.3.7) we use that after convergence of the value vector u_i we have by (10.3.3) and (10.3.4)

$$\tilde{g}_k^* - \tilde{r}_k(s, \tilde{\pi}_k(s)) = \sum_{s'} \tilde{P}_k(s'|s, \tilde{\pi}_k(s)) \cdot u_i(s') - u_i(s).$$

Then noting that $v_k(s,a) = 0$ for $a \neq \tilde{\pi}_k(s)$ and using vector/matrix notation it follows that

$$\begin{aligned}&\sum_{s,a} v_k(s,a)\big(\tilde{g}_k^* - \tilde{r}_k(s, \tilde{\pi}_k(s))\big) \\ &= \sum_{s,a} v_k(s,a)\Big(\sum_{s'} \tilde{P}_k(s'|s, \tilde{\pi}_k(s)) \cdot u_i(s') - u_i(s)\Big) \\ &= \mathbf{v}_k\big(\tilde{\mathbf{P}}_k - \mathbf{I}\big)\mathbf{u} \\ &= \mathbf{v}_k\big(\tilde{\mathbf{P}}_k - \mathbf{P}_k + \mathbf{P}_k - \mathbf{I}\big)\mathbf{w}_k \\ &= \mathbf{v}_k\big(\tilde{\mathbf{P}}_k - \mathbf{P}_k\big)\mathbf{w}_k + \mathbf{v}_k\big(\mathbf{P}_k - \mathbf{I}\big)\mathbf{w}_k, \quad (10.3.9)\end{aligned}$$

where $\mathbf{P}_k$ is the true transition matrix (in μ) of the optimistic policy $\tilde{\pi}_k$ in episode k, and $\mathbf{w}_k$ is a renormalization of the vector $\mathbf{u}$ (with entries $u_i(s)$) where $w_k(s) := u_i(s) - \frac{1}{2}(\min_s u_i(s) + \max_s u_i(s))$, so that $\|\mathbf{w}_k\|_\infty \le \frac{D}{2}$ by Lemma 10.3.4.

Since $\|\tilde{\mathbf{P}}_k - \mathbf{P}_k\|_1 \le \|\tilde{\mathbf{P}}_k - \hat{\mathbf{P}}_k\|_1 + \|\hat{\mathbf{P}}_k - \mathbf{P}_k\|_1$, the first term of (10.3.9) is bounded as

$$\begin{aligned} \mathbf{v}_k(\tilde{\mathbf{P}}_k - \mathbf{P}_k)\mathbf{w}_k &\le \left\|\mathbf{v}_k(\tilde{\mathbf{P}}_k - \mathbf{P}_k)\right\|_1 \cdot \left\|\mathbf{w}_k\right\|_\infty \\ &\le 2\sum_{s,a} v_k(s,a)\,\mathrm{conf}_k^p(s,a)\,D. \end{aligned} \tag{10.3.10}$$

The second term can be rewritten as martingale difference sequence

$$\begin{aligned} \mathbf{v}_k(\mathbf{P}_k - \mathbf{I})\mathbf{w}_k &= \sum_{t=t_k}^{t_{k+1}-1}\Big(P(\cdot|s_t,a)\mathbf{w}_k - w_k(s_t)\Big) \\ &= \sum_{t=t_k}^{t_{k+1}-1}\Big(P(\cdot|s_t,a)\mathbf{w}_k - w_k(s_{t+1})\Big) + w_k(s_{t_{k+1}}) - w_k(s_{t_k}), \end{aligned}$$

so that its sum over all episodes can be bounded by Azuma-Hoeffding inequality (5.11) and Lemma 10.3.2, that is,

$$\sum_k \mathbf{v}_k(\mathbf{P}_k - \mathbf{I})\mathbf{w}_k \le D\sqrt{\tfrac{5}{2}T\log\left(\tfrac{8T}{\delta}\right)} + DSA\log_2\left(\tfrac{8T}{SA}\right) \tag{10.3.11}$$

with probability at least $1 - \frac{\delta}{12T^{5/4}}$.

Summing (10.3.8) and (10.3.10) over all episodes, by definition of the confidence intervals and Lemma 10.3.3 we have

$$\begin{aligned} &\sum_k\sum_{s,a} v_k(s,a)\,\mathrm{conf}_k^r(s,a) + 2D\sum_k\sum_{s,a} v_k(s,a)\,\mathrm{conf}_k^p(s,a) \\ &\le \mathrm{const}\cdot D\sqrt{S\log(AT/\delta)}\sum_k\sum_{s,a}\tfrac{v_k(s,a)}{\sqrt{N_k(s,a)}} \\ &\le \mathrm{const}\cdot D\sqrt{S\log(AT/\delta)}\sqrt{SAT}. \end{aligned} \tag{10.3.12}$$

Thus, combining (10.3.7)–(10.3.12) we obtain that

$$\sum_{s,a} v_k(s,a)\big(\tilde{g}_k^* - r(s,a)\big) \le \mathrm{const}\cdot D\sqrt{S\log(AT/\delta)}\sqrt{SAT} \tag{10.3.13}$$

with probability at least $1 - \frac{\delta}{12T^{5/4}}$.

Finally by (10.3.6) and (10.3.13) the regret of UCRL2 is upper bounded by $\mathrm{const}\cdot D\sqrt{S\log(AT/\delta)}\sqrt{SAT}$ with probability at least $1 - 3\sum_{T\ge 2}\frac{\delta}{12T^{5/4}} \ge 1-\delta$. □

The following is a corresponding lower bound on the regret that shows that the upper bound of Theorem 10.3.1 is optimal in T and A.

Theorem 10.3.2 (Jaksch et al. [12]) *For any algorithm and any natural numbers T, S, $A > 1$, and $D \geq \log_A S$ there is an MDP with S states, A actions, and diameter D, the expected regret after T steps is*

$$\Omega\big(\sqrt{DSAT}\big).$$

Similar to the distribution dependent regret bound of Theorem 10.2.1 for UCB1, one can derive a logarithmic bound on the expected regret of UCRL2.

Theorem 10.3.3 (Jaksch et al. [12]) *In an MDP with S states, A actions, and diameter D the expected regret of UCRL2 is*

$$O\left(\frac{D^2 S^2 A \log(T)}{\Delta}\right),$$

where $\Delta \triangleq g^* - \max_\pi \left\{g^\pi : g^\pi < g^*\right\}$ is the gap between the optimal gain and the second largest gain.

10.3.2 Bibliographical Remarks

Similar to UCB1 that was based on the work of Lai and Robbins [3], UCRL2 is not the first optimistic algorithm with theoretical guarantees. Thus, the *index policies* of Burnetas and Katehakis [14] and Tewari and Bartlett [15] choose actions optimistically by using confidence bounds for the estimates in the current state. However, the logarithmic regret bounds are derived only for *ergodic* MDPs in which each policy visits each state with probability 1.

Another important predecessor that is based on the principle of *optimism in the face of uncertainty* is R-Max [16], that assumes in each not sufficiently visited state to receive the maximal possible reward. UCRL2 offers a refinement of this idea to motivate exploration. Sample complexity bounds as derived for R-Max can also be obtained for UCRL2, cf. [12].

The gap between the lower bound of Theorem 10.3.2 and the bound for UCRL2 has not been closed so far. There have been various tries in that direction for different algorithms inspired by Thompson sampling [17] or UCB1 [18]. However all of the claimed proofs seem to contain some issues that remain unresolved up-to-date.

The situation is settled in the simpler episodic setting, where after any H steps there is a restart. Here there are matching upper and lower bounds of order $\sqrt{HSAT}$ on the regret, see [19].

In the discounted setting, the MBIE algorithm of Strehl and Littman [20, 21] is a precursor of UCRL2 that is based on the same ideas. While there are regret bounds available also for MBIE, these are not easily comparable to Theorem 10.2.1, as the regret is measured along the trajectory of the algorithm, while the regret considered for UCRL2 is with respect to the trajectory an optimal policy would have taken. In general, regret in the discounted setting seems to be a less satisfactory concept. However, sample complexity bounds in the discounted setting for a UCRL2 variant have been given in Lattimore and Hutter [22].

Last but not least, we would like to refer any reader interested in the material of this chapter to the recent book of Lattimore and Szepesvári [10] that deals with the whole range of topics from simple bandits to reinforcement learning in MDPs in much more detail.

References

1. Hart, P.E., Nilsson, N.J., Raphael, B.: A formal basis for the heuristic determination of minimum cost paths. IEEE Trans. Syst. Sci. Cybern. **4**(2), 100–107 (1968)
2. Auer, P., Cesa-Bianchi, N., Fischer, P.: Finite time analysis of the multiarmed bandit problem. Mach. Learn. **47**(2–3), 235–256 (2002)
3. Tze Leung Lai and Herbert Robbins: Asymptotically efficient adaptive allocation rules. Adv. Appl. Math. **6**(1), 4–22 (1985)
4. Mannor, S., Tsitsiklis, J.N.: The sample complexity of exploration in the multi-armed bandit problem. J. Mach. Learn. Res. **5**, 623–648 (2004)
5. Auer, P., Ortner, R.: UCB revisited: improved regret bounds for the stochastic multi-armed bandit problem. Period. Math. Hung. **61**(1–2), 55–65 (2010)
6. Lattimore, T.: Optimally confident UCB: Improved regret for finite-armed bandits. Technical Report 1507.07880 (2015). (arXiv)
7. Audibert, J.-Y., Bubeck, S.: Minimax policies for adversarial and stochastic bandits. In: colt2009. Proceedings of the 22nd Annual Conference on Learning Theory, pp. 217–226 (2009)
8. Bubeck, S., Cesa-Bianchi, N.: Regret analysis of stochastic and nonstochastic multi-armed bandit problems. Found. Trends Mach. Learn. **5**(1), 1–122 (2012)
9. Auer, P., Cesa-Bianchi, N., Freund, Y., Schapire, R.E.: The nonstochastic multiarmed bandit problem. SIAM J. Comput. **32**(1), 48–77 (2002)
10. Lattimore, T., Szepesvári, C.: Bandit Algorithms. Cambridge University Press (2020)
11. Ortner, R., Ryabko, D., Auer, P., Munos, R.: Regret bounds for restless Markov bandits. Theor. Comput. Sci. **558**, 62–76 (2014)
12. Jaksch, T., Ortner, R., Auer, P.: Near-optimal regret bounds for reinforcement learning. J. Mach. Learn. Res. **11**, 1563–1600 (2010)
13. Weissman, T., Ordentlich, E., Seroussi, G., Verdu, S., Weinberger, M.J.: Inequalities for the L_1 deviation of the empirical distribution. Technical Report HPL-2003-97 (R.1), Hewlett-Packard Labs, Technical Report (2003)
14. Burnetas, A.N., Katehakis, M.N.: Optimal adaptive policies for Markov decision processes. Math. Oper. Res. **22**(1), 222–255 (1997)
15. Tewari, A., Bartlett, P.: Optimistic linear programming gives logarithmic regret for irreducible MDPs. In: Advances in Neural Information Processing Systems, vol. 20, pp. 1505–1512. MIT Press (2008)
16. Brafman, R.I., Tennenholtz, M.: R-MAX-A general polynomial time algorithm for near-optimal reinforcement learning. J. Mach. Learn. Res. **3**, 213–231 (2003)

17. Agrawal, S., Jia, R.: Optimistic posterior sampling for reinforcement learning: worst-case regret bounds. In: Advances in Neural Information Processing Systems, vol. 30, pp. 1184–1194 (2017)
18. Ortner, R.: Regret bounds for reinforcement learning via Markov chain concentration. J. Artif. Intell. Res. **67**, 115–128 (2020)
19. Azar, M.G., Osband, I., Munos, R.: Minimax regret bounds for reinforcement learning. In: Proceedings of the 34th International Conference on Machine Learning, ICML 2017, pp. 263–272 (2017)
20. Strehl, A.L., Littman, M.L.: A theoretical analysis of model-based interval estimation. In: Machine Learning, Proceedings of the 22nd International Conference, ICML 2005, pp. 857–864. ACM (2005)
21. Strehl, A.L., Littman, M.L.: An analysis of model-based interval estimation for Markov decision processes. J. Comput. Syst. Sci. **74**(8), 1309–1331 (2008)
22. Lattimore, T., Hutter, M.: Near-optimal PAC bounds for discounted MDPs. Theor. Comput. Sci. **558**, 125–143 (2014)

Chapter 11
Conclusion

This book touched upon the basic principles of decision making under uncertainty in the context of reinforcement learning. While one of the main streams of thought is Bayesian decision theory, we also discussed the basics of approximate dynamic programming and stochastic approximation as applied to reinforcement learning problems.

Consciously, however, we have avoided going into a number of topics related to reinforcement learning and decision theory, some of which would need a book of their own to be properly addressed. Even though it was fun writing the book, we at some point had to decide to stop and consolidate the material we had, sometimes culling partially developed material in favour of a more concise volume.

Firstly, we haven't explicitly considered many models that can be used for representing transition distributions, value functions or policies, beyond the simplest ones, as we felt that this would detract from the main body of the text. Textbooks for the latest fashion are always going to be abundant, and we hope that this book provides a sufficient basis to enable the use of any current methods. There are also a large number of areas which have not been covered at all. In particular, while we touched upon the setting of two-player games and its connection to robust statistical decisions, we have not examined problems which are also relevant to sequential decision making, such as Markov games and Bayesian games. In relation to this, while early in the book we discuss risk aversion and risk seeking, we have not discussed specific sequential decision making algorithms for such problems. Furthermore, even though we discuss the problem of preference elicitation, we do not discuss specific algorithms for it or the related problem of inverse reinforcement learning. Another topic which went unmentioned, but which may become more important in the future, is hierarchical reinforcement learning as well as options, which allow constructing long-term actions (such as "go to the supermarket") from primitive actions (such as "open the door"). Finally, even though we have mentioned the basic framework of

C. Dimitrakakis and R. Ortner, *Decision Making Under Uncertainty and Reinforcement Learning*, Intelligent Systems Reference Library 223,
https://doi.org/10.1007/978-3-031-07614-5_11

regret minimization, we focused on the standard reinforcement learning problem, and ignored adversarial settings and problems with varying amounts of side information.

It is important to note that the book almost entirely elides social aspects of decision making. In practice, any algorithm that is going to be used to make autonomous decision is going to have a societal impact. In such cases, the algorithm designer must guard against negative externalities, such as hurting disadvantaged groups, violating privacy, or environmental damage. However, as a lot of these issues are context dependent, we urge the reader to consult recent work in economics, algorithmic fairness and differential privacy.

Appendix
Symbols

Table A.1 Logic symbols

$\triangleq$	Definition
$\wedge$	Logical and
$\vee$	Logical or
$\Rightarrow$	Implies
$\Leftrightarrow$	If and only if
$\exists$	There exists
$\forall$	For every
s.t.	Such that

Table A.2 List of set theory symbols

$\{x_k\}$	A set indexed by k
$\{x \mid xRy\}$	The set of x satisfying relation xRy
$\mathbb{N}$	Set of natural numbers
$\mathbb{Z}$	Set of integers
$\mathbb{R}$	Set of real numbers
Ω	The universe set (or sample space)
$\emptyset$	The empty set
$\mathbb{\Delta}^n$	The n-dimensional simplex
$\mathbb{\Delta}(A)$	The collection of distributions over a set A
$\mathfrak{B}(A)$	The Borel σ-algebra induced by a set A

(continued)

C. Dimitrakakis and R. Ortner, *Decision Making Under Uncertainty and Reinforcement Learning*, Intelligent Systems Reference Library 223,
https://doi.org/10.1007/978-3-031-07614-5

Table A.2 (continued)

A^n	The product set $\prod_{i=1}^{n} A$
A^*	$\bigcup_{n=0}^{\infty} A^n$ the set of all sequences from set A
$x \in A$	x belongs to A
$A \subset B$	A is a (strict) subset of B
$A \subseteq B$	A is a (non-strict) subset of B
$B \setminus A$	Set difference
$B \triangle A$	Symmetric set difference
$A^{\complement}$	Set complement
$A \cup B$	Set union
$A \cap B$	Set intersection

Table A.3 Analysis and linear algebra symbols

$\boldsymbol{x}^{\top}$	The transpose of a vector $\boldsymbol{x}$
$\lvert \boldsymbol{A} \rvert$	The determinant of a matrix A
$\lVert x \rVert_p$	The p-norm of a vector $(\sum_i \lvert x_i \rvert^p)^{1/p}$
$\lVert f \rVert_p$	The p-norm of a function $(\int \lvert f(x) i \rvert^p \, \mathrm{d}x)^{1/p}$
$\lVert A \rVert_p$	The operator norm of a matrix $\max \left\{ Ax \mid \lVert x \rVert_p = 1 \right\}$
$\partial f(x)/\partial x_i$	Partial derivative with respect to x_i
∇f	Gradient vector of partial derivatives with respect to vector x

Table A.4 Miscellaneous statistics symbols

$\mathcal{N}(\mu, m)$	Normal distribution with mean μ and covariance Σ
$\mathcal{N}(\mu, \Sigma)$	Normal distribution with mean μ and covariance Σ
$\mathcal{Bern}(\omega)$	Bernoulli distribution with parameter ω
$\mathcal{Binom}(\omega, t)$	Binomial distribution with parameter ω over t trials
$\mathcal{Gamma}(\alpha, \beta)$	Gamma distribution with shape α and scaling β
$\mathcal{Dir}(\boldsymbol{\alpha})$	Dirichlet distribution with prior mass $\boldsymbol{\alpha}$
$\mathcal{Unif}(A)$	Uniform distribution on the set A
$\mathcal{Beta}(\alpha, \beta)$	Beta distribution with parameters (α, β)
$\mathcal{Geom}(\omega)$	Geometric distribution with parameter ω
$\mathcal{Wish}(n-1, \boldsymbol{T})$	Wishart distribution with n degrees of freedom and parameter matrix $\boldsymbol{T}$
$\phi : \mathcal{X} \to \mathcal{Y}$	Statistic mapping from observations to a vector space

Index

C. Dimitrakakis and R. Ortner, *Decision Making Under Uncertainty and Reinforcement Learning*, Intelligent Systems Reference Library 223,
https://doi.org/10.1007/978-3-031-07614-5